Electrical Engineering: Concepts and Applications

Electrical Engineering: Concepts and Applications

Edited by
Evan Ryan

Larsen & Keller
www.larsen-keller.com

Electrical Engineering: Concepts and Applications
Edited by Evan Ryan
ISBN: 978-1-63549-098-5 (Hardback)

 Larsen & Keller

Published by Larsen and Keller Education,
5 Penn Plaza,
19th Floor,
New York, NY 10001, USA

Cataloging-in-Publication Data

Electrical engineering : concepts and applications / edited by Evan Ryan.
    p. cm.
Includes bibliographical references and index.
ISBN 978-1-63549-098-5
1. Electrical engineering. 2. Electric power. 3. Electric currents. I. Ryan, Evan.
TK146 .E44 2017
621.3--dc23

The publisher's policy is to use permanent paper from mills that operate a sustainable forestry policy. Furthermore, the publisher ensures that the text paper and cover boards used have met acceptable environmental accreditation standards.

Printed and bound in the United States of America.

For more information regarding Larsen and Keller Education and its products, please visit the publisher's website www.larsen-keller.com

Table of Contents

Preface

Electrical engineering is a branch of engineering which deals with the study of electromagnetism, electricity and electronics. It is an overarching field and includes various subfields like radio-frequency engineering, power engineering, signal processing, telecommunications, microelectronics, digital computers, control systems, instrumentation and radio-frequency engineering, etc. This book aims to provide detailed information about this vast subject. It gives thorough insights into the various subfields of this area. It is a valuable compilation of topics, ranging from the basic to the most complex theories and principles in the field of electrical engineering. Such selected concepts that redefine this field have been presented in this textbook. Different approaches, evaluations and methodologies on electrical engineering have been included in it. This textbook is meant for students who are looking for an elaborate reference text on this field.

A detailed account of the significant topics covered in this book is provided below:

Chapter 1- The branch of electrical engineering that studies electronics and electricity is known as electrical engineering. Modern technology has helped electronics in becoming a part of daily life and in becoming a necessity. This chapter will provide an integrated understanding of electrical engineering.

Chapter 2- Electrical engineering has numerous subdiciplines. Some of these branches are power engineering, control engineering, electronic engineering, telecommunication engineering, electromechanics etc. Power engineering is the engineering of the circulation of electricity and it includes devices such as transformers, electric generators and power electronics. This section is a compilation of the various branches of electrical engineering that forms an integral part of the broader subject matter.

Chapter 3- Electricity, electronics and electromagnetism are important topics related to the field of electrical engineering. Electronics controls electrical energy, it is concerned with components such as vacuum tubes and integrated circuits. The following chapter unfolds its key elements in a critical yet systematic manner.

Chapter 4- In order to completely understand electrical engineering, it is necessary to understand the principles and concepts of the subject. Electrical wiring is necessary for devices such as meters and light fittings whereas a daisy chain is a wiring system which has numerous gadgets connected to it in a sequence. The following section elucidates the mechanisms associated with this area of study.

Chapter 5- The physical laws related to electrical engineering are discussed in this chapter. Some of these laws are Ampère's circuital law, Kirchhoff's circuit laws, Coulomb's law and Ohm's law. This chapter helps the readers in understanding the laws related to electrical engineering.

Chapter 6- Electric power is the degree of energy being transmitted by electric circuits. Electric current is the movement of electric charge. Other topics covered in this chapter are electric machine, passive sign convention, AC power, power station etc. This section has been carefully written to provide an easy understanding of electric power and current.

Chapter 7- Volt is the unit used in electrical engineering, and ampere is the SI unit of electric current. The other measurement units elucidated in this text are ohm, farad and henry. This section is an overview of the subject matter incorporating all the major aspects of the measurement units of electrical engineering.

Chapter 8- A number of applications are based on electrical engineering. Some of these are avionics, transistor, voltmeter, electrical telegraph and electric motor. Avionics are used in satellites and spacecrafts and it helps in the navigation and management of the systems found in these vehicles. The aspects elucidated in this chapter are of vital importance, and provide a better understanding of electrical engineering.

Chapter 9- This section focuses on the progress made by electrical engineering in the last few decades. It includes the span from the discovery of electrical current to the harnessing of electric energy. The theme of this chapter helps the readers in developing an in-depth understanding of electrical engineering.

I would like to make a special mention of my publisher who considered me worthy of this opportunity and also supported me throughout the process. I would also like to thank the editing team at the back-end who extended their help whenever required.

Editor

Introduction to Electrical Engineering

The branch of electrical engineering that studies electronics and electricity is known as electrical engineering. Modern technology has helped electronics in becoming a part of daily life and in becoming a necessity. This chapter will provide an integrated understanding of electrical engineering.

Electrical engineering is a field of engineering that generally deals with the study and application of electricity, electronics, and electromagnetism. This field first became an identifiable occupation in the latter half of the 19th century after commercialization of the electric telegraph, the telephone, and electric power distribution and use. Subsequently, broadcasting and recording media made electronics part of daily life. The invention of the transistor, and later the integrated circuit, brought down the cost of electronics to the point they can be used in almost any household object.

Electrical engineers design complex power systems on a macroscopic level as well as microscopic electronic devices

Electrical engineering has now subdivided into a wide range of subfields including electronics, digital computers, power engineering, telecommunications, control systems, radio-frequency engineering, signal processing, instrumentation, and microelectronics. The subject of electronic engineering is often treated as its own subfield but it intersects with all the other subfields, including the power electronics of power engineering.

Electrical engineers typically hold a degree in electrical engineering or electronic engineering. Practicing engineers may have professional certification and be members of a professional body. Such bodies include the Institute of Electrical and Electronics Engineers (IEEE) and the Institution of Engineering and Technology (professional society) (IET).

Electrical engineers work in a very wide range of industries and the skills required are likewise variable. These range from basic circuit theory to the management skills required of a project manager. The tools and equipment that an individual engineer may need are similarly variable, ranging from a simple voltmeter to a top end analyzer to sophisticated design and manufacturing software.

and electronic circuitry, which recently achieved the record setting length of 1 nanometer for a single gate.

History

Electricity has been a subject of scientific interest since at least the early 17th century. A prominent early electrical scientist was William Gilbert who was the first to draw a clear distinction between magnetism and static electricity and is credited with establishing the term electricity. He also designed the versorium: a device that detected the presence of statically charged objects. Then in 1762 Swedish professor Johan Carl Wilcke invented, and in 1775 Alessandro Volta improved, a device (for which Volta coined the name electrophorus) that produced a static electric charge, and by 1800 Volta had developed the voltaic pile, a forerunner of the electric battery.

19th Century

The discoveries of Michael Faraday formed the foundation of electric motor technology

In 19th century, research into the subject started to intensify. Notable developments in this century include the work of Georg Ohm, who in 1827 quantified the relationship between the electric current and potential difference in a conductor, of Michael Faraday, the discoverer of electromagnetic induction in 1831, and of James Clerk Maxwell, who in 1873 published a unified theory of electricity and magnetism in his treatise *Electricity and Magnetism*.

Electrical engineering became a profession in the later 19th century. Practitioners had created a global electric telegraph network and the first professional electrical engineering institutions were founded in the UK and USA to support the new discipline. Although it is impossible to precisely pinpoint a first electrical engineer, Francis Ronalds stands ahead of the field, who created the first working electric telegraph system in 1816 and documented his vision of how the world could be transformed by electricity. Over 50 years later, he joined the new Society of Telegraph Engineers (soon to be renamed the Institution of Electrical Engineers) where he was regarded by other members as the first of their cohort. By the end of the 19th century, the world had been forever changed by the rapid communication made possible by the engineering development of land-lines, submarine cables, and, from about 1890, wireless telegraphy.

Practical applications and advances in such fields created an increasing need for standardised units of measure. They led to the international standardization of the units volt, ampere, coulomb, ohm, farad, and henry. This was achieved at an international conference in Chicago in 1893. The publication of these standards formed the basis of future advances in standardisation in various industries, and in many countries the definitions were immediately recognised in relevant legislation.

During these years, the study of electricity was largely considered to be a subfield of physics. That's because early electrical technology was electromechanical in nature. The Technische Universität Darmstadt founded the world's first department of electrical engineering in 1882. The first electrical engineering degree program was started at Massachusetts Institute of Technology (MIT) in the physics department under Professor Charles Cross, though it was Cornell University to produce the world's first electrical engineering graduates in 1885. The first course in electrical engineering was taught in 1883 in Cornell's Sibley College of Mechanical Engineering and Mechanic Arts. It was not until about 1885 that Cornell President Andrew Dickson White established the first Department of Electrical Engineering in the United States. In the same year, University College London founded the first chair of electrical engineering in Great Britain. Professor Mendell P. Weinbach at University of Missouri soon followed suit by establishing the electrical engineering department in 1886. Afterwards, universities and institutes of technology gradually started to offer electrical engineering programs to their students all over the world.

Thomas Edison, electric light and (DC) power supply networks

Károly Zipernowsky, Ottó Bláthy, Miksa Déri, the ZDB transformer

William Stanley, Jr., transformers

Galileo Ferraris, electrical theory, induction motor

Nikola Tesla, practical polyphase (AC) and induction motor designs

Mikhail Dolivo-Dobrovolsky developed standard 3-phase (AC) systems

Charles Proteus Steinmetz, AC mathematical theories for engineers

Oliver Heaviside, developed theoretical models for electric circuits

During these decades use of electrical engineering increased dramatically. In 1882, Thomas Edison switched on the world's first large-scale electric power network that provided 110 volts — direct current (DC) — to 59 customers on Manhattan Island in New York City. In 1884, Sir Charles Parsons invented the steam turbine allowing for more efficient electric power generation. Alternating current, with its ability to transmit power more efficiently over long distances via the use of transformers, developed rapidly in the 1880s and 1890s with transformer designs by Károly Zipernowsky, Ottó Bláthy and Miksa Déri (later called ZBD transformers), Lucien Gaulard, John Dixon Gibbs and William Stanley, Jr.. Practical AC motor designs including induction motors were independently invented by Galileo Ferraris and Nikola Tesla and further developed into a practical three-phase form by Mikhail Dolivo-Dobrovolsky and Charles Eugene Lancelot Brown. Charles Steinmetz and Oliver Heaviside contributed to the theoretical basis of alternating current engineering. The spread in the use of AC set off in the United States what has been called the *War of Currents* between a George Westinghouse backed AC system and a Thomas Edison backed DC power system, with AC being adopted as the overall standard.

More Modern Developments

Guglielmo Marconi known for his pioneering work on long distance radio transmission

During the development of radio, many scientists and inventors contributed to radio technology and electronics. The mathematical work of James Clerk Maxwell during the 1850s had shown the relationship of different forms of electromagnetic radiation including possibility of invisible airborne waves (later called "radio waves"). In his classic physics experiments of 1888, Heinrich Hertz proved Maxwell's theory by transmitting radio waves with a spark-gap transmitter, and detected them by using simple electrical devices. Other physicists experimented with these new waves and in the process developed devices for transmitting and detecting them. In 1895, Guglielmo Marconi began work on a way to adapt the known methods of transmitting and detecting these "Hertzian waves" into a purpose built commercial wireless telegraphic system. Early on, he sent wireless signals over a distance of one and a half miles. In December 1901, he sent wireless waves that were not affected by the curvature of the Earth. Marconi later transmitted the wireless signals across the Atlantic between Poldhu, Cornwall, and St. John's, Newfoundland, a distance of 2,100 miles (3,400 km).

In 1897, Karl Ferdinand Braun introduced the cathode ray tube as part of an oscilloscope, a crucial enabling technology for electronic television. John Fleming invented the first radio tube, the diode, in 1904. Two years later, Robert von Lieben and Lee De Forest independently developed the amplifier tube, called the triode.

In 1920, Albert Hull developed the magnetron which would eventually lead to the development of the microwave oven in 1946 by Percy Spencer. In 1934, the British military began to make strides toward radar (which also uses the magnetron) under the direction of Dr Wimperis, culminating in the operation of the first radar station at Bawdsey in August 1936.

A replica of the first working transistor.

In 1941, Konrad Zuse presented the Z3, the world's first fully functional and programmable computer using electromechanical parts. In 1943, Tommy Flowers designed and built the Colossus, the world's first fully functional, electronic, digital and programmable computer. In 1946, the ENIAC (Electronic Numerical Integrator and Computer) of John Presper Eckert and John Mauchly followed, beginning the computing era. The arithmetic performance of these machines allowed engineers to develop completely new technologies and achieve new objectives, including the Apollo program which culminated in landing astronauts on the Moon.

Solid-state Transistors

The invention of the transistor in late 1947 by William B. Shockley, John Bardeen, and Walter Brattain of the Bell Telephone Laboratories opened the door for more compact devices and led to the development of the integrated circuit in 1958 by Jack Kilby and independently in 1959 by Robert Noyce. Starting in 1968, Ted Hoff and a team at the Intel Corporation invented the first commercial microprocessor, which foreshadowed the personal computer. The Intel 4004 was a four-bit processor released in 1971, but in 1973 the Intel 8080, an eight-bit processor, made the first personal computer, the Altair 8800, possible.

Subdisciplines

Electrical engineering has many subdisciplines, the most common of which are listed below. Although there are electrical engineers who focus exclusively on one of these subdisciplines, many deal with a combination of them. Sometimes certain fields, such as electronic engineering and computer engineering, are considered separate disciplines in their own right.

Power

Power pole

Power engineering deals with the generation, transmission, and distribution of electricity as well as the design of a range of related devices. These include transformers, electric generators, electric motors, high voltage engineering, and power electronics. In many regions of the world, governments maintain an electrical network called a power grid that connects a variety of generators together with users of their energy. Users purchase electrical energy from the grid, avoiding the costly exercise of having to generate their own. Power engineers may work on the design and maintenance of the power grid as well as the power systems that connect to it. Such systems are called *on-grid* power systems and may supply the grid with additional power, draw power from the grid, or do both. Power engineers may also work on systems that do not connect to the grid, called *off-grid* power systems, which in some cases are preferable to on-grid systems. The future includes Satellite controlled power systems, with feedback in real time to prevent power surges and prevent blackouts.

Control

Control systems play a critical role in space flight.

Control engineering focuses on the modeling of a diverse range of dynamic systems and the design of controllers that will cause these systems to behave in the desired manner. To implement such controllers, electrical engineers may use electronic circuits, digital signal processors, microcontrollers, and programmable logic controls (PLCs). Control engineering has a wide range of applications from the flight and propulsion systems of commercial airliners to the cruise control present in many modern automobiles. It also plays an important role in industrial automation.

Control engineers often utilize feedback when designing control systems. For example, in an automobile with cruise control the vehicle's speed is continuously monitored and fed back to the system which adjusts the motor's power output accordingly. Where there is regular feedback, control theory can be used to determine how the system responds to such feedback.

Electronics

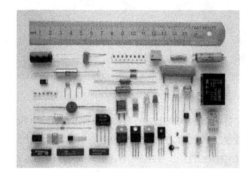

Electronic components

Electronic engineering involves the design and testing of electronic circuits that use the properties of components such as resistors, capacitors, inductors, diodes, and transistors to achieve a particular functionality. The tuned circuit, which allows the user of a radio to filter out all but a single station, is just one example of such a circuit. Another example (of a pneumatic signal conditioner) is shown in the adjacent photograph.

Prior to the Second World War, the subject was commonly known as *radio engineering* and basically was restricted to aspects of communications and radar, commercial radio, and early television. Later, in post war years, as consumer devices began to be developed, the field grew to include modern television, audio systems, computers, and microprocessors. In the mid-to-late 1950s, the term *radio engineering* gradually gave way to the name *electronic engineering*.

Before the invention of the integrated circuit in 1959, electronic circuits were constructed from discrete components that could be manipulated by humans. These discrete circuits consumed much space and power and were limited in speed, although they are still common in some applications. By contrast, integrated circuits packed a large number—often millions—of tiny electrical components, mainly transistors, into a small chip around the size of a coin. This allowed for the powerful computers and other electronic devices we see today.

Microelectronics

Microelectronics engineering deals with the design and microfabrication of very small electronic circuit components for use in an integrated circuit or sometimes for use on their own as a general electronic component. The most common microelectronic components are semiconductor transistors, although all main electronic components (resistors, capacitors etc.) can be created at a microscopic level. Nanoelectronics is the further scaling of devices down to nanometer levels. Modern devices are already in the nanometer regime, with below 100 nm processing having been standard since about 2002.

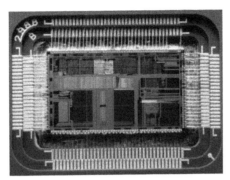

Microprocessor

Microelectronic components are created by chemically fabricating wafers of semiconductors such as silicon (at higher frequencies, compound semiconductors like gallium arsenide and indium phosphide) to obtain the desired transport of electronic charge and control of current. The field of microelectronics involves a significant amount of chemistry and material science and requires the electronic engineer working in the field to have a very good working knowledge of the effects of quantum mechanics.

Signal Processing

A Bayer filter on a CCD requires signal processing to get a red, green, and blue value at each pixel.

Signal processing deals with the analysis and manipulation of signals. Signals can be either analog, in which case the signal varies continuously according to the information, or digital, in which case the signal varies according to a series of discrete values representing the information. For analog signals, signal processing may involve the amplification and filtering of audio signals for audio equipment or the modulation and demodulation of signals for telecommunications. For digital signals, signal processing may involve the compression, error detection and error correction of digitally sampled signals.

Signal Processing is a very mathematically oriented and intensive area forming the core of digital signal processing and it is rapidly expanding with new applications in every field of electrical engineering such as communications, control, radar, audio engineering, broadcast engineering, power electronics, and biomedical engineering as many already existing analog systems are replaced with their digital counterparts. Analog signal processing is still important in the design of many control systems.

DSP processor ICs are found in every type of modern electronic systems and products including, SDTV | HDTV sets, radios and mobile communication devices, Hi-Fi audio equipment, Dolby noise reduction algorithms, GSM mobile phones, mp3 multimedia players, camcorders and digital cameras, automobile control systems, noise cancelling headphones, digital spectrum analyzers, intelligent missile guidance, radar, GPS based cruise control systems, and all kinds of image processing, video processing, audio processing, and speech processing systems.

Telecommunications

Satellite dishes are a crucial component in the analysis of satellite information.

Telecommunications engineering focuses on the transmission of information across a channel such as a coax cable, optical fiber or free space. Transmissions across free space require information to be encoded in a carrier signal to shift the information to a carrier frequency suitable for transmission; this is known as modulation. Popular analog modulation techniques include amplitude modulation and frequency modulation. The choice of modulation affects the cost and performance of a system and these two factors must be balanced carefully by the engineer.

Once the transmission characteristics of a system are determined, telecommunication engineers design the transmitters and receivers needed for such systems. These two are sometimes combined to form a two-way communication device known as a transceiver. A key consideration in the design of transmitters is their power consumption as this is closely related to their signal strength. If the signal strength of a transmitter is insufficient the signal's information will be corrupted by noise.

Instrumentation

Instrumentation engineering deals with the design of devices to measure physical quantities such as pressure, flow, and temperature. The design of such instrumentation requires a good understanding of physics that often extends beyond electromagnetic theory. For example, flight instruments measure variables such as wind speed and altitude to enable pilots the control of aircraft analytically. Similarly, thermocouples use the Peltier-Seebeck effect to measure the temperature difference between two points.

Flight instruments provide pilots with the tools to control aircraft analytically.

Often instrumentation is not used by itself, but instead as the sensors of larger electrical systems. For example, a thermocouple might be used to help ensure a furnace's temperature remains constant. For this reason, instrumentation engineering is often viewed as the counterpart of control engineering.

Computers

Supercomputers are used in fields as diverse as computational biology and geographic information systems.

Computer engineering deals with the design of computers and computer systems. This may involve the design of new hardware, the design of PDAs, tablets, and supercomputers, or the use of computers to control an industrial plant. Computer engineers may also work on a system's software. However, the design of complex software systems is often the domain of software engineering, which is usually considered a separate discipline. Desktop computers represent a tiny fraction of the devices a computer engineer might work on, as computer-like architectures are now found in a range of devices including video game consoles and DVD players.

Related Disciplines

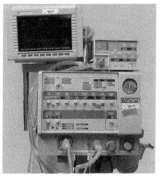

The Bird VIP Infant ventilator

Mechatronics is an engineering discipline which deals with the convergence of electrical and mechanical systems. Such combined systems are known as electromechanical systems and have widespread adoption. Examples include automated manufacturing systems, heating, ventilation and air-conditioning systems, and various subsystems of aircraft and automobiles.

The term *mechatronics* is typically used to refer to macroscopic systems but futurists have predicted the emergence of very small electromechanical devices. Already, such small devices, known as Microelectromechanical systems (MEMS), are used in automobiles to tell airbags when to deploy, in digital projectors to create sharper images, and in inkjet printers to create nozzles for high definition printing. In the future it is hoped the devices will help build tiny implantable medical devices and improve optical communication.

Biomedical engineering is another related discipline, concerned with the design of medical equipment. This includes fixed equipment such as ventilators, MRI scanners, and electrocardiograph monitors as well as mobile equipment such as cochlear implants, artificial pacemakers, and artificial hearts.

Aerospace engineering and robotics an example is the most recent electric propulsion and ion propulsion.

Education

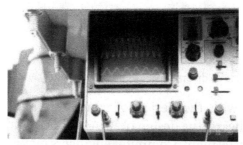

Oscilloscope

Electrical engineers typically possess an academic degree with a major in electrical engineering, electronics engineering, electrical engineering technology, or electrical and electronic engineering. The same fundamental principles are taught in all programs, though emphasis may vary according to title. The length of study for such a degree is usually four or five years and the completed degree may be designated as a Bachelor of Science in Electrical/Electronics Engineering Technology, Bachelor of Engineering, Bachelor of Science, Bachelor of Technology, or Bachelor of Applied Science depending on the university. The bachelor's degree generally includes units covering physics, mathematics, computer science, project management, and a variety of topics in electrical engineering. Initially such topics cover most, if not all, of the subdisciplines of electrical engineering. At some schools, the students can then choose to emphasize one or more subdisciplines towards the end of their courses of study.

At many schools, electronic engineering is included as part of an electrical award, sometimes explicitly, such as a Bachelor of Engineering (Electrical and Electronic), but in others electrical and electronic engineering are both considered to be sufficiently broad and complex that separate degrees are offered.

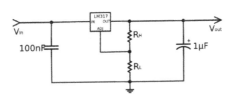

Typical electrical engineering diagram used as a troubleshooting tool

Some electrical engineers choose to study for a postgraduate degree such as a Master of Engineering/Master of Science (M.Eng./M.Sc.), a Master of Engineering Management, a Doctor of Philosophy (Ph.D.) in Engineering, an Engineering Doctorate (Eng.D.), or an Engineer's degree. The master's and engineer's degrees may consist of either research, coursework or a mixture of the two. The Doctor of Philosophy and Engineering Doctorate degrees consist of a significant research component and are often viewed as the entry point to academia. In the United Kingdom and some other European countries, Master of Engineering is often considered to be an undergraduate degree of slightly longer duration than the Bachelor of Engineering rather than postgraduate.

Practicing Engineers

Belgian electrical engineers inspecting the rotor of a 40,000 kilowatt turbine of the General Electric Company in New York City

In most countries, a bachelor's degree in engineering represents the first step towards professional certification and the degree program itself is certified by a professional body. After completing a certified degree program the engineer must satisfy a range of requirements (including work experience requirements) before being certified. Once certified the engineer is designated the title of Professional Engineer (in the United States, Canada and South Africa), Chartered Engineer or Incorporated Engineer (in India, Pakistan, the United Kingdom, Ireland and Zimbabwe), Chartered Professional Engineer (in Australia and New Zealand) or European Engineer (in much of the European Union).

The IEEE corporate office is on the 17th floor of 3 Park Avenue in New York City

The advantages of certification vary depending upon location. For example, in the United States and Canada "only a licensed engineer may seal engineering work for public and private clients".

This requirement is enforced by state and provincial legislation such as Quebec's Engineers Act. In other countries, no such legislation exists. Practically all certifying bodies maintain a code of ethics that they expect all members to abide by or risk expulsion. In this way these organizations play an important role in maintaining ethical standards for the profession. Even in jurisdictions where certification has little or no legal bearing on work, engineers are subject to contract law. In cases where an engineer's work fails he or she may be subject to the tort of negligence and, in extreme cases, the charge of criminal negligence. An engineer's work must also comply with numerous other rules and regulations such as building codes and legislation pertaining to environmental law.

Professional bodies of note for electrical engineers include the Institute of Electrical and Electronics Engineers (IEEE) and the Institution of Engineering and Technology (IET). The IEEE claims to produce 30% of the world's literature in electrical engineering, has over 360,000 members worldwide and holds over 3,000 conferences annually. The IET publishes 21 journals, has a worldwide membership of over 150,000, and claims to be the largest professional engineering society in Europe. Obsolescence of technical skills is a serious concern for electrical engineers. Membership and participation in technical societies, regular reviews of periodicals in the field and a habit of continued learning are therefore essential to maintaining proficiency. An MIET(Member of the Institution of Engineering and Technology) is recognised in Europe as an Electrical and computer (technology) engineer.

In Australia, Canada, and the United States electrical engineers make up around 0.25% of the labor force.

Tools and Work

From the Global Positioning System to electric power generation, electrical engineers have contributed to the development of a wide range of technologies. They design, develop, test, and supervise the deployment of electrical systems and electronic devices. For example, they may work on the design of telecommunication systems, the operation of electric power stations, the lighting and wiring of buildings, the design of household appliances, or the electrical control of industrial machinery.

Satellite communications is typical of what electrical engineers work on.

Fundamental to the discipline are the sciences of physics and mathematics as these help to obtain both a qualitative and quantitative description of how such systems will work. Today most engineering work involves the use of computers and it is commonplace to use computer-aided design programs when designing electrical systems. Nevertheless, the ability to sketch ideas is still invaluable for quickly communicating with others.

The Shadow robot hand system

Although most electrical engineers will understand basic circuit theory (that is the interactions of elements such as resistors, capacitors, diodes, transistors, and inductors in a circuit), the theories employed by engineers generally depend upon the work they do. For example, quantum mechanics and solid state physics might be relevant to an engineer working on VLSI (the design of integrated circuits), but are largely irrelevant to engineers working with macroscopic electrical systems. Even circuit theory may not be relevant to a person designing telecommunication systems that use off-the-shelf components. Perhaps the most important technical skills for electrical engineers are reflected in university programs, which emphasize strong numerical skills, computer literacy, and the ability to understand the technical language and concepts that relate to electrical engineering.

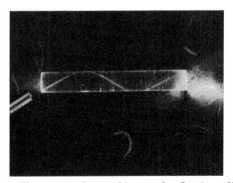

A laser bouncing down an acrylic rod, illustrating the total internal reflection of light in a multi-mode optical fiber.

A wide range of instrumentation is used by electrical engineers. For simple control circuits and alarms, a basic multimeter measuring voltage, current, and resistance may suffice. Where time-varying signals need to be studied, the oscilloscope is also an ubiquitous instrument. In RF engineering and high frequency telecommunications, spectrum analyzers and network analyzers are used. In some disciplines, safety can be a particular concern with instrumentation. For instance, medical electronics designers must take into account that much lower voltages than normal can be dangerous when electrodes are directly in contact with internal body fluids. Power transmission engineering also has great safety concerns due to the high voltages used; although voltmeters may in principle be similar to their low voltage equivalents, safety and calibration issues make them very different. Many disciplines of electrical engineering use tests specific to their discipline. Audio electronics engineers use audio test sets consisting of a signal generator and a meter, principally to measure level but also other parameters such as harmonic distortion and noise. Likewise, information technology have their own test sets, often specific to a particular data format, and the same is true of television broadcasting.

Radome at the Misawa Air Base Misawa Security Operations Center, Misawa, Japan

For many engineers, technical work accounts for only a fraction of the work they do. A lot of time may also be spent on tasks such as discussing proposals with clients, preparing budgets and determining project schedules. Many senior engineers manage a team of technicians or other engineers and for this reason project management skills are important. Most engineering projects involve some form of documentation and strong written communication skills are therefore very important.

The workplaces of engineers are just as varied as the types of work they do. Electrical engineers may be found in the pristine lab environment of a fabrication plant, the offices of a consulting firm or on site at a mine. During their working life, electrical engineers may find themselves supervising a wide range of individuals including scientists, electricians, computer programmers, and other engineers.

Electrical engineering has an intimate relationship with the physical sciences. For instance, the physicist Lord Kelvin played a major role in the engineering of the first transatlantic telegraph cable. Conversely, the engineer Oliver Heaviside produced major work on the mathematics of transmission on telegraph cables. Electrical engineers are often required on major science projects. For instance, large particle accelerators such as CERN need electrical engineers to deal with many aspects of the project: from the power distribution, to the instrumentation, to the manufacture and installation of the superconducting electromagnets.

References

- Ronalds, B.F. (2016). Sir Francis Ronalds: Father of the Electric Telegraph. London: Imperial College Press. ISBN 978-1-78326-917-4.

- Engineering: Issues, Challenges and Opportunities for Development. UNESCO. 2010. pp. 127–8. ISBN 978-92-3-104156-3.

- Manual on the Use of Thermocouples in Temperature Measurement. ASTM International. 1 January 1993. p. 154. ISBN 978-0-8031-1466-1.

- Occupational Outlook Handbook, 2008–2009. U S Department of Labor, Jist Works. 1 March 2008. p. 148. ISBN 978-1-59357-513-7.

Branches of Electrical Engineering

Electrical engineering has numerous subdiciplines. Some of these branches are power engineering, control engineering, electronic engineering, telecommunication engineering, electromechanics etc. Power engineering is the engineering of the circulation of electricity and it includes devices such as transformers, electric generators and power electronics. This section is a compilation of the various branches of electrical engineering that forms an integral part of the broader subject matter.

Power Engineering

A steam turbine used to provide electric power.

Power engineering, also called power systems engineering, is a subfield of energy engineering and electrical engineering that deals with the generation, transmission, distribution and utilization of electric power and the electrical devices connected to such systems including generators, motors and transformers. Although much of the field is concerned with the problems of three-phase AC power – the standard for large-scale power transmission and distribution across the modern world – a significant fraction of the field is concerned with the conversion between AC and DC power and the development of specialized power systems such as those used in aircraft or for electric railway networks. Power engineering draws the majority of its theoretical base from electrical engineering and while some power engineers could be considered energy engineers, energy engineers often do not have the theoretical electrical engineering background to understand power engineering.

History

Electricity became a subject of scientific interest in the late 17th century with the work of William Gilbert. Over the next two centuries a number of important discoveries were made including the incandescent light bulb and the voltaic pile. Probably the greatest discovery with respect to power engineering came from Michael Faraday who in 1831 discovered that a change in magnetic flux induces an electromotive force in a loop of wire—a principle known as electromagnetic induction that helps explain how generators and transformers work.

A sketch of the Pearl Street Station, the first steam-powered electric power station in New York City

In 1881 two electricians built the world's first power station at Godalming in England. The station employed two waterwheels to produce an alternating current that was used to supply seven Siemens arc lamps at 250 volts and thirty-four incandescent lamps at 40 volts. However supply was intermittent and in 1882 Thomas Edison and his company, The Edison Electric Light Company, developed the first steam-powered electric power station on Pearl Street in New York City. The Pearl Street Station consisted of several generators and initially powered around 3,000 lamps for 59 customers. The power station used direct current and operated at a single voltage. Since the direct current power could not be easily transformed to the higher voltages necessary to minimise power loss during transmission, the possible distance between the generators and load was limited to around half-a-mile (800 m).

That same year in London Lucien Gaulard and John Dixon Gibbs demonstrated the first transformer suitable for use in a real power system. The practical value of Gaulard and Gibbs' transformer was demonstrated in 1884 at Turin where the transformer was used to light up forty kilometres (25 miles) of railway from a single alternating current generator. Despite the success of the system, the pair made some fundamental mistakes. Perhaps the most serious was connecting the primaries of the transformers in series so that switching one lamp on or off would affect other lamps further down the line. Following the demonstration George Westinghouse, an American entrepreneur, imported a number of the transformers along with a Siemens generator and set his engineers to experimenting with them in the hopes of improving them for use in a commercial power system.

One of Westinghouse's engineers, William Stanley, recognised the problem with connecting transformers in series as opposed to parallel and also realised that making the iron core of a transformer a fully enclosed loop would improve the voltage regulation of the secondary winding. Using this knowledge he built the worlds first practical transformer based alternating current power system at Great Barrington, Massachusetts in 1886. In 1885 the Italian physicist and electrical engineer Galileo Ferraris demonstrated an induction motor and in 1887 and 1888 the Serbian-American engineer Nikola Tesla filed a range of patents related to power systems including one for a practical two-phase induction motor which Westinghouse licensed for his AC system.

By 1890 the power industry had flourished and power companies had built thousands of power systems (both direct and alternating current) in the United States and Europe – these networks were effectively dedicated to providing electric lighting. During this time a fierce rivalry in the US known as the "War of Currents" emerged between Edison and Westinghouse over which form of transmission (direct or alternating current) was superior. In 1891, Westinghouse installed the first major power system that was designed to drive an electric motor and not just provide electric lighting. The installation powered a 100 horsepower (75 kW) synchronous motor at Telluride, Colorado with the motor being started by a Tesla induction motor. On the other side of the Atlantic, Oskar von Miller built a 20 kV 176 km three-phase transmission line from Lauffen am Neckar to Frankfurt am Main for the Electrical Engineering Exhibition in Frankfurt. In 1895, after a protracted decision-making process, the Adams No. 1 generating station at Niagara Falls began transmitting three-phase alternating current power to Buffalo at 11 kV. Following completion of the Niagara Falls project, new power systems increasingly chose alternating current as opposed to direct current for electrical transmission.

Although the 1880s and 1890s were seminal decades in the field, developments in power engineering continued throughout the 20th and 21st century. In 1936 the first commercial high-voltage direct current (HVDC) line using mercury-arc valves was built between Schenectady and Mechanicville, New York. HVDC had previously been achieved by installing direct current generators in series (a system known as the Thury system) although this suffered from serious reliability issues. In 1957 Siemens demonstrated the first solid-state rectifier (solid-state rectifiers are now the standard for HVDC systems) however it was not until the early 1970s that this technology was used in commercial power systems. In 1959 Westinghouse demonstrated the first circuit breaker that used SF_6 as the interrupting medium. SF_6 is a far superior dielectric to air and, in recent times, its use has been extended to produce far more compact switching equipment (known as switchgear) and transformers. Many important developments also came from extending innovations in the ICT field to the power engineering field. For example, the development of computers meant load flow studies could be run more efficiently allowing for much better planning of power systems. Advances in information technology and telecommunication also allowed for much better remote control of the power system's switchgear and generators.

Basics of Electric Power

An external AC to DC power adapter used for household appliances

Electric power is the mathematical product of two quantities: current and voltage. These two quantities can vary with respect to time (AC power) or can be kept at constant levels (DC power).

Most refrigerators, air conditioners, pumps and industrial machinery use AC power whereas computers and digital equipment use DC power (the digital devices you plug into the mains typically have an internal or external power adapter to convert from AC to DC power). AC power has the advantage of being easy to transform between voltages and is able to be generated and utilised by brushless machinery. DC power remains the only practical choice in digital systems and can be more economical to transmit over long distances at very high voltages.

The ability to easily transform the voltage of AC power is important for two reasons: Firstly, power can be transmitted over long distances with less loss at higher voltages. So in power networks where generation is distant from the load, it is desirable to step-up the voltage of power at the generation point and then step-down the voltage near the load. Secondly, it is often more economical to install turbines that produce higher voltages than would be used by most appliances, so the ability to easily transform voltages means this mismatch between voltages can be easily managed.

Solid state devices, which are products of the semiconductor revolution, make it possible to transform DC power to different voltages, build brushless DC machines and convert between AC and DC power. Nevertheless, devices utilising solid state technology are often more expensive than their traditional counterparts, so AC power remains in widespread use.

Power

Transmission lines transmit power across the grid.

Power Engineering deals with the generation, transmission, distribution and utilization of electricity as well as the design of a range of related devices. These include transformers, electric generators, electric motors and power electronics.

The power grid is an electrical network that connects a variety of electric generators to the users of electric power. Users purchase electricity from the grid so that they do not need to generate their own. Power engineers may work on the design and maintenance of the power grid as well as the power systems that connect to it. Such systems are called on-grid power systems and may supply the grid with additional power, draw power from the grid or do both. The grid is designed and managed using software that performs simulations of power flows.

Power engineers may also work on systems that do not connect to the grid. These systems are called off-grid power systems and may be used in preference to on-grid systems for a variety of reasons. For example, in remote locations it may be cheaper for a mine to generate its own power rather than pay for connection to the grid and in most mobile applications connection to the grid is simply not practical.

Today, most grids adopt three-phase electric power with alternating current. This choice can be partly attributed to the ease with which this type of power can be generated, transformed and used. Often (especially in the United States), the power is split before it reaches residential customers whose low-power appliances rely upon single-phase electric power. However, many larger industries and organizations still prefer to receive the three-phase power directly because it can be used to drive highly efficient electric motors such as three-phase induction motors.

Transformers play an important role in power transmission because they allow power to be converted to and from higher voltages. This is important because higher voltages suffer less power loss during transmission. This is because higher voltages allow for lower current to deliver the same amount of power, as power is the product of the two. Thus, as the voltage steps up, the current steps down. It is the current flowing through the components that result in both the losses and the subsequent heating. These losses, appearing in the form of heat, are equal to the current squared times the electrical resistance through which the current flows, so as the voltage goes up the losses are dramatically reduced.

For these reasons, electrical substations exist throughout power grids to convert power to higher voltages before transmission and to lower voltages suitable for appliances after transmission.

Components

Power engineering is a network of interconnected components which convert different forms of energy to electrical energy. Modern power engineering consists of four main subsystems: the generation subsystem, the transmission subsystem, the distribution subsystem and the utilization subsystem. In the generation subsystem, the power plant produces the electricity. The transmission subsystem transmits the electricity to the load centers. The distribution subsystem continues to transmit the power to the customers. The utilization system is concerned with the different uses of electrical energy like illumination, refrigeration, traction, electric drives, etc. Utilization is a very recent concept in Power engineering.

Generation

Generation of electrical power is a process whereby energy is transformed into an electrical form. There are several different transformation processes, among which are chemical, photo-voltaic, and electromechanical. Electromechanical energy conversion is used in converting energy from coal, petroleum, natural gas, uranium, or water flow into electrical energy. Of these, all except the wind energy conversion process take advantage of the synchronous AC generator coupled to a steam, gas or hydro turbine such that the turbine converts steam, gas, or water flow into rotational energy, and the synchronous generator then converts the rotational energy of the turbine into electrical energy. It is the turbine-generator conversion process that is by far most economical and consequently most common in the industry today.

The AC synchronous machine is the most common technology for generating electrical energy. It is called synchronous because the composite magnetic field produced by the three stator windings rotate at the same speed as the magnetic field produced by the field winding on the rotor. A simplified circuit model is used to analyze steady-state operating conditions for a synchronous machine. The phasor diagram is an effective tool for visualizing the relationships between internal voltage, armature current, and terminal voltage. The excitation control system is used on synchronous machines to regulate terminal voltage, and the turbine-governor system is used to regulate the speed of the machine. However, in highly interconnected systems, such as the "Western system", the "Texas system" and the "Eastern system", *one* machine will usually be assigned as the so-called "swing machine", and which generation may be increased or decreased to compensate for small changes in load, thereby maintaining the system frequency at precisely 60 Hz. Should the load dramatically change, as happens with a system separation, then a combination of "spinning reserve" and the "swing machine" may be used by the system's load dispatcher.

The operating costs of generating electrical energy is determined by the fuel cost and the efficiency of the power station. The efficiency depends on generation level and can be obtained from the heat rate curve. The incremental cost curve may also be derived from the heat rate curve. *Economic dispatch* is the process of allocating the required load demand between the available generation units such that the cost of operation is minimized. *Emission dispatch* is the process of allocating the required load demand between the available generation units such that air pollution occurring from operation is minimized. In large systems, particularly in the West, a combination of economic and emission dispatch may be used.

Transmission

The electricity is transported to load locations from a power station to a transmission subsystem. Therefore, we may think of the transmission system as providing the medium of transportation for electric energy. The transmission system may be subdivided into the bulk transmission system and the sub-transmission system. The functions of the bulk transmission are to interconnect generators, to interconnect various areas of the network, and to transfer electrical energy from the generators to the major load centers. This portion of the system is called "bulk" because it delivers energy only to so-called bulk loads such as the distribution system of a town, city, or large industrial plant. The function of the sub-transmission system is to interconnect the bulk power system with the distribution system.

Transmission circuits may be built either underground or overhead. Underground cables are used predominantly in urban areas where acquisition of overhead rights of way are costly or not possible. They are also used for transmission under rivers, lakes and bays. Overhead transmission is used otherwise because, for a given voltage level, overhead conductors are much less expensive than underground cables.

The transmission system is a highly integrated system. It is referred to as the substation equipment and transmission lines. The substation equipment contain the transformers, relays, and circuit breakers. Transformers are important static devices which transfer electrical energy from one circuit to another in the transmission subsystem. Transformers are used to step up the voltage on the transmission line to reduce the power loss which is dissipated on the way. A relay is functionally a level-detector; they perform a switching action when the input voltage (or current) meets or exceeds a specific and adjustable value. A circuit breaker is an automatically operated electrical switch designed to protect an electrical circuit from damage caused by overload or short circuit. A change in the status of any one component can significantly affect the operation of the entire system. Without adequate contact protection, the occurrence of undesired electric arcing causes significant degradation of the contacts, which suffer serious damage. There are three possible causes for power flow limitations to a transmission line. These causes are thermal overload, voltage instability, and rotor angle instability. Thermal overload is caused by excessive current flow in a circuit causing overheating. Voltage instability is said to occur when the power required to maintain voltages at or above acceptable levels exceeds the available power. Rotor angle instability is a dynamic problem that may occur following faults, such as short circuit, in the transmission system. It may also occur tens of seconds after a fault due to poorly damped or undamped oscillatory response of the rotor motion. As long as the *equal area criteria* is maintained, the interconnected system will remain stable. Should the *equal area criteria* be violated, it becomes necessary to separate the unstable component from the remainder of the system.

Distribution

The distribution system transports the power from the transmission system/substation to the customer. Distribution feeders can be radial or networked in an open loop configuration with a single or multiple alternate sources. Rural systems tend to be of the former and urban systems the latter. The equipment associated with the distribution system usually begins downstream of the distribution feeder circuit breaker. The transformer and circuit breaker are usually under the jurisdiction of a "substations department". The distribution feeders consist of combinations of overhead and underground conductor, 3 phase and single phase switches with load break and non-loadbreak ability, relayed protective devices, fuses, transformers (to utilization voltage), surge arresters, voltage regulators and capacitors.

More recently, Smart Grid initiatives are being deployed so that: One, distribution feeder faults are automatically isolated and power restored to unfaulted circuits by automatic hardware/software/communications packages. And two, capacitors are automatically switched on or off to dynamically control VAR flow and for CVR (Conservation Voltage Reduction).

Utilization

Utilization is the "end result" of the generation, transmission, and distribution of electric power. The energy carried by the transmission and distribution system is turned into useful work, light, heat, or a combination of these items at the utilization point. Understanding and characterizing the utilization of electric power is critical for proper planning and operation of power systems. Improper characterization of utilization can result of over or under building of power system facilities and stressing of system equipment beyond design capabilities. The term load refers to a device or collection of devices that draw energy from the power system. Individual loads (devices) range from small light bulbs to large induction motors to arc furnaces. The term load is often somewhat arbitrarily applied, at times being used to describe a specific device, and other times referring to an entire facility and even being used to describe the lumped power requirements of power system components and connected utilization devices downstream of a specific point in large scale system studies.

A major application of electric energy is in its conversion to mechanical energy. Electromagnetic, or "EM" devices designed for this purpose are commonly called "motors." Actually the machine is the central component of an integrated system consisting of the source, controller, motor, and load. For specialized applications, the system may be, and frequently is, designed as an integrated whole. Many household appliances (e.g., a vacuum cleaner) have in one unit, the controller, the motor, and the load.

Control Engineering

Control systems play a critical role in space flight

Control engineering or control systems engineering is the engineering discipline that applies control theory to design systems with desired behaviors. The practice uses sensors to measure the output performance of the device being controlled and those measurements can be used to give feedback to the input actuators that can make corrections toward desired performance. When a device is designed to perform without the need of human inputs for correction it is called automatic control (such as cruise control for regulating the speed of a car). Multi-disciplinary in nature, control systems engineering activities focus on implementation of control systems mainly derived by mathematical modeling of systems of a diverse range.

Overview

Modern day control engineering is a relatively new field of study that gained significant attention during the 20th century with the advancement of technology. It can be broadly defined or classified as practical application of control theory. Control engineering has an essential role in a wide range of control systems, from simple household washing machines to high-performance F-16 fighter aircraft. It seeks to understand physical systems, using mathematical modeling, in terms of inputs, outputs and various components with different behaviors, use control systems design tools to develop controllers for those systems and implement controllers in physical systems employing available technology. A system can be mechanical, electrical, fluid, chemical, financial and even biological, and the mathematical modeling, analysis and controller design uses control theory in one or many of the time, frequency and complex-s domains, depending on the nature of the design problem.

History

Automatic control systems were first developed over two thousand years ago. The first feedback control device on record is thought to be the ancient Ktesibios's water clock in Alexandria, Egypt around the third century B.C. It kept time by regulating the water level in a vessel and, therefore, the water flow from that vessel. This certainly was a successful device as water clocks of similar design were still being made in Baghdad when the Mongols captured the city in 1258 A.D. A variety of automatic devices have been used over the centuries to accomplish useful tasks or simply to just entertain. The latter includes the automata, popular in Europe in the 17th and 18th centuries, featuring dancing figures that would repeat the same task over and over again; these automata are examples of open-loop control. Milestones among feedback, or "closed-loop" automatic control devices, include the temperature regulator of a furnace attributed to Drebbel, circa 1620, and the centrifugal flyball governor used for regulating the speed of steam engines by James Watt in 1788.

In his 1868 paper "On Governors", James Clerk Maxwell was able to explain instabilities exhibited by the flyball governor using differential equations to describe the control system. This demonstrated the importance and usefulness of mathematical models and methods in understanding complex phenomena, and signaled the beginning of mathematical control and systems theory. Elements of control theory had appeared earlier but not as dramatically and convincingly as in Maxwell's analysis.

Control theory made significant strides in the next 100 years. New mathematical techniques made it possible to control, more accurately, significantly more complex dynamical systems than the original flyball governor. These techniques include developments in optimal control in the 1950s and 1960s, followed by progress in stochastic, robust, adaptive and optimal control methods in the 1970s and 1980s. Applications of control methodology have helped make possible space travel and communication satellites, safer and more efficient aircraft, cleaner auto engines, cleaner and more efficient chemical processes.

Before it emerged as a unique discipline, control engineering was practiced as a part of mechanical engineering and control theory was studied as a part of electrical engineering since electrical circuits can often be easily described using control theory techniques. In the very first control relationships, a current output was represented with a voltage control input. However, not having proper technology to implement electrical control systems, designers left with the option of less efficient and slow responding mechanical systems. A very effective mechanical controller that is still widely used in some hydro plants is the governor. Later on, previous to modern power electronics, process control systems for industrial applications were devised by mechanical engineers using pneumatic and hydraulic control devices, many of which are still in use today.

Control Theory

There are two major divisions in control theory, namely, classical and modern, which have direct implications over the control engineering applications. The scope of classical control theory is limited to single-input and single-output (SISO) system design, except when analyzing for disturbance rejection using a second input. The system analysis is carried out in the time domain using differential equations, in the complex-s domain with the Laplace transform, or in the frequency domain by transforming from the complex-s domain. Many systems may be assumed to have a second order and single variable system response in the time domain. A controller designed using classical theory often requires on-site tuning due to incorrect design approximations. Yet, due to the easier physical implementation of classical controller designs as compared to systems designed using modern control theory, these controllers are preferred in most industrial applications. The most common controllers designed using classical control theory are PID controllers. A less common implementation may include either or both a Lead or Lag filter. The ultimate end goal is to meet a requirements set typically provided in the time-domain called the Step response, or at times in the frequency domain called the Open-Loop response. The Step response characteristics applied in a specification are typically percent overshoot, settling time, etc. The Open-Loop response characteristics applied in a specification are typically Gain and Phase margin and bandwidth. These characteristics may be evaluated through simulation including a dynamic model of the system under control coupled with the compensation model.

In contrast, modern control theory is carried out in the state space, and can deal with multiple-input and multiple-output (MIMO) systems. This overcomes the limitations of classical control theory in more sophisticated design problems, such as fighter aircraft control, with the limitation that no frequency domain analysis is possible. In modern design, a system is represented to the greatest advantage as a set of decoupled first order differential equations defined using state variables. Nonlinear, multivariable, adaptive and robust control theories come under this division. Matrix methods are significantly limited for MIMO systems where linear independence cannot be assured in the relationship between inputs and outputs. Being fairly new, modern control theory has many areas yet to be explored. Scholars like Rudolf E. Kalman and Aleksandr Lyapunov are well-known among the people who have shaped modern control theory.

Control Systems

Control engineering is the engineering discipline that focuses on the modeling of a diverse range of dynamic systems (e.g. mechanical systems) and the design of controllers that will cause these systems to behave in the desired manner. Although such controllers need not be electrical many are and hence control engineering is often viewed as a subfield of electrical engineering. However, the falling price of microprocessors is making the actual implementation of a control system essentially trivial. As a result, focus is shifting back to the mechanical and process engineering discipline, as intimate knowledge of the physical system being controlled is often desired.

Electrical circuits, digital signal processors and microcontrollers can all be used to implement control systems. Control engineering has a wide range of applications from the flight and propulsion systems of commercial airliners to the cruise control present in many modern automobiles.

In most of the cases, control engineers utilize feedback when designing control systems. This is often accomplished using a PID controller system. For example, in an automobile with cruise control the vehicle's speed is continuously monitored and fed back to the system, which adjusts the motor's torque accordingly. Where there is regular feedback, control theory can be used to determine how the system responds to such feedback. In practically all such systems stability is important and control theory can help ensure stability is achieved.

Although feedback is an important aspect of control engineering, control engineers may also work on the control of systems without feedback. This is known as open loop control. A classic example of open loop control is a washing machine that runs through a pre-determined cycle without the use of sensors.

Control Engineering Education

At many universities, control engineering courses are taught in electrical and electronic engineering, mechatronics engineering, mechanical engineering, and aerospace engineering. In others, control engineering is connected to computer science, as most control techniques today are implemented through computers, often as embedded systems (as in the automotive field). The field of control within chemical engineering is often known as process control. It deals primarily with the control of variables in a chemical process in a plant. It is taught as part of the undergraduate curriculum of any chemical engineering program and employs many of the same principles in control engineering. Other engineering disciplines also overlap with control engineering as it can be applied to any system for which a suitable model can be derived. However, specialised control engineering departments do exist, for example, the Department of Automatic Control and Systems Engineering at the University of Sheffield and the Department of Systems Engineering at the United States Naval Academy.

Control engineering has diversified applications that include science, finance management, and even human behavior. Students of control engineering may start with a linear control system course dealing with the time and complex-s domain, which requires a thorough background in elementary mathematics and Laplace transform, called classical control theory. In linear control, the student does frequency and time domain analysis. Digital control and nonlinear control courses require Z transformation and algebra respectively, and could be said to complete a basic control education.

Recent Advancement

Originally, control engineering was all about continuous systems. Development of computer control tools posed a requirement of discrete control system engineering because the communications between the computer-based digital controller and the physical system are governed by a computer clock. The equivalent to Laplace transform in the discrete domain is the Z-transform. Today, many of the control systems are computer controlled and they consist of both digital and analog components.

Therefore, at the design stage either digital components are mapped into the continuous domain and the design is carried out in the continuous domain, or analog components are mapped into discrete domain and design is carried out there. The first of these two methods is more commonly encountered in practice because many industrial systems have many continuous systems components, including mechanical, fluid, biological and analog electrical components, with a few digital controllers.

Similarly, the design technique has progressed from paper-and-ruler based manual design to computer-aided design and now to computer-automated design or CAutoD which has been made possible by evolutionary computation. CAutoD can be applied not just to tuning a predefined control scheme, but also to controller structure optimisation, system identification and invention of novel control systems, based purely upon a performance requirement, independent of any specific control scheme.

Resilient Control Systems extends the traditional focus on addressing only plant disturbances to frameworks, architectures and methods that address multiple types of unexpected disturbance. In particular, adapting and transforming behaviors of the control system in response to malicious actors, abnormal failure modes, undesirable human action, etc. Development of resilience technologies require the involvement of multidisciplinary teams to holistically address the performance challenges.

Electronic Engineering

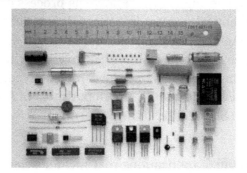

Electronic components

Printed circuit board

Electronics engineering, or electronic engineering, is an engineering discipline which utilizes non-linear and active electrical components (such as semiconductor devices, especially transistors, diodes and integrated circuits) to design electronic circuits, devices, Microprocessors/Microcontrollers and systems including VHDL Modelling for Programmable logic devices and FPGAs. The discipline typically also designs passive electrical components, usually based on printed circuit boards.

The term "electronic engineering" denotes a broad engineering field that covers subfields such as analog electronics, digital electronics, consumer electronics, embedded systems and power electronics. Electronics engineering deals with implementation of applications, principles and algorithms developed within many related fields, for example solid-state physics, radio engineering, telecommunications, control systems, signal processing, systems engineering, computer engineering, instrumentation engineering, electric power control, robotics, and many others.

The Institute of Electrical and Electronics Engineers (IEEE) is one of the most important and influential organizations for electronics engineers.

Relationship to Electrical Engineering

Electronics is a subfield within the wider electrical engineering academic subject. An academic degree with a major in electronics engineering can be acquired from some universities, while other universities use electrical engineering as the subject. The term electrical engineer is still used in the academic world to include electronic engineers. However, some people consider the term 'electrical engineer' should be reserved for those having specialized in power and heavy current or high voltage engineering, while others consider that power is just one subset of electrical engineering and (and indeed the term 'power engineering' is used in that industry) as well as 'electrical distribution engineering'. Again, in recent years there has been a growth of new separate-entry degree courses such as 'information engineering', 'systems engineering' and 'communication systems engineering', often followed by academic departments of similar name, which are typically not considered as subfields of electronics engineering but of electrical engineering.

History

Electronic engineering as a profession sprang from technological improvements in the telegraph industry in the late 19th century and the radio and the telephone industries in the early 20th century. People were attracted to radio by the technical fascination it inspired, first in receiving and then in transmitting. Many who went into broadcasting in the 1920s were only 'amateurs' in the period before World War I.

To a large extent, the modern discipline of electronic engineering was born out of telephone, radio, and television equipment development and the large amount of electronic systems development during World War II of radar, sonar, communication systems, and advanced munitions and weapon systems. In the interwar years, the subject was known as radio engineering and it was only in the late 1950s that the term electronic engineering started to emerge.

Electronics

In the field of electronic engineering, engineers design and test circuits that use the electromagnetic properties of electrical components such as resistors, capacitors, inductors, diodes and transistors to achieve a particular functionality. The tuner circuit, which allows the user of a radio to filter out all but a single station, is just one example of such a circuit.

In designing an integrated circuit, electronics engineers first construct circuit schematics that specify the electrical components and describe the interconnections between them. When completed, VLSI engineers convert the schematics into actual layouts, which map the layers of various conductor and semiconductor materials needed to construct the circuit. The conversion from schematics to layouts can be done by software but very often requires human fine-tuning to decrease space and power consumption. Once the layout is complete, it can be sent to a fabrication plant for manufacturing.

Integrated circuits and other electrical components can then be assembled on printed circuit boards to form more complicated circuits. Today, printed circuit boards are found in most electronic devices including televisions, computers and audio players.

Subfields

Electronic engineering has many subfields. This section describes some of the most popular subfields in electronic engineering; although there are engineers who focus exclusively on one subfield, there are also many who focus on a combination of subfields.

Signal processing deals with the analysis and manipulation of signals. Signals can be either analog, in which case the signal varies continuously according to the information, or digital, in which case the signal varies according to a series of discrete values representing the information.

For analog signals, signal processing may involve the amplification and filtering of audio signals for audio equipment or the modulation and demodulation of signals for telecommunications. For digital signals, signal processing may involve the compression, error checking and error detection of digital signals.

Telecommunications engineering deals with the transmission of information across a channel such as a co-axial cable, optical fiber or free space.

Transmissions across free space require information to be encoded in a carrier wave in order to shift the information to a carrier frequency suitable for transmission, this is known as modulation. Popular analog modulation techniques include amplitude modulation and frequency modulation. The choice of modulation affects the cost and performance of a system and these two factors must be balanced carefully by the engineer.

Once the transmission characteristics of a system are determined, telecommunication engineers design the transmitters and receivers needed for such systems. These two are sometimes combined to form a two-way communication device known as a transceiver. A key consideration in the design of transmitters is their power consumption as this is closely related to their signal strength. If the signal strength of a transmitter is insufficient the signal's information will be corrupted by noise.

Control engineering has a wide range of applications from the flight and propulsion systems of commercial airplanes to the cruise control present in many modern cars. It also plays an important role in industrial automation.

Control engineers often utilize feedback when designing control systems. For example, in a car with cruise control the vehicle's speed is continuously monitored and fed back to the system which adjusts the engine's power output accordingly. Where there is regular feedback, control theory can be used to determine how the system responds to such feedback.

Instrumentation engineering deals with the design of devices to measure physical quantities such as pressure, flow and temperature. These devices are known as instrumentation.

The design of such instrumentation requires a good understanding of physics that often extends beyond electromagnetic theory. For example, radar guns use the Doppler effect to measure the speed of oncoming vehicles. Similarly, thermocouples use the Peltier–Seebeck effect to measure the temperature difference between two points.

Often instrumentation is not used by itself, but instead as the sensors of larger electrical systems. For example, a thermocouple might be used to help ensure a furnace's temperature remains constant. For this reason, instrumentation engineering is often viewed as the counterpart of control engineering.

Computer engineering deals with the design of computers and computer systems. This may involve the design of new computer hardware, the design of PDAs or the use of computers to control an industrial plant. Development of embedded systems—systems made for specific tasks (e.g., mobile phones)—is also included in this field. This field includes the micro controller and its applications. Computer engineers may also work on a system's software. However, the design of complex software systems is often the domain of software engineering, which is usually considered a separate discipline.

VLSI design engineering VLSI stands for *very large scale integration*. It deals with fabrication of ICs and various electronics components.

Typical Undergraduate Syllabus

Apart from electromagnetics and network theory, other items in the syllabus are particular to *electronics* engineering course. *Electrical* engineering courses have other specialisms such as machines, power generation and distribution. This list does not include the extensive engineering mathematics curriculum that is a prerequisite to a degree.

Electromagnetics

Elements of vector calculus: divergence and curl; Gauss' and Stokes' theorems, Maxwell's equations: differential and integral forms. Wave equation, Poynting vector. Plane waves: propagation through various media; reflection and refraction; phase and group velocity; skin depth. Transmission lines: characteristic impedance; impedance transformation; Smith chart; impedance matching; pulse excitation. Waveguides: modes in rectangular waveguides; boundary conditions; cut-off frequencies; dispersion relations. Antennas: Dipole antennas; antenna arrays; radiation pattern; reciprocity theorem, antenna gain.

Network Analysis

Network graphs: matrices associated with graphs; incidence, fundamental cut set and fundamental circuit matrices. Solution methods: nodal and mesh analysis. Network theorems: superposition, Thevenin and Norton's maximum power transfer, Wye-Delta transformation. Steady state sinusoidal analysis using phasors. Linear constant coefficient differential equations; time domain analysis of simple RLC circuits, Solution of network equations using Laplace transform: frequency domain analysis of RLC circuits. 2-port network parameters: driving point and transfer functions. State equations for networks.

Electronic Devices and Circuits

Electronic devices: Energy bands in silicon, intrinsic and extrinsic silicon. Carrier transport in silicon: diffusion current, drift current, mobility, resistivity. Generation and recombination of carriers. p-n junction diode, Zener diode, tunnel diode, BJT, JFET, MOS capacitor, MOSFET, LED, p-i-n and avalanche photo diode, LASERs. Device technology: integrated circuit fabrication process, oxidation, diffusion, ion implantation, photolithography, n-tub, p-tub and twin-tub CMOS process.

Analog circuits: Equivalent circuits (large and small-signal) of diodes, BJTs, JFETs, and MOS-FETs. Simple diode circuits, clipping, clamping, rectifier. Biasing and bias stability of transistor and FET amplifiers. Amplifiers: single-and multi-stage, differential, operational, feedback and power. Analysis of amplifiers; frequency response of amplifiers. Simple op-amp circuits. Filters. Sinusoidal oscillators; criterion for oscillation; single-transistor and op-amp configurations. Function generators and wave-shaping circuits, Power supplies.

Digital circuits: Boolean functions (NOT, AND, OR, XOR,...). Logic gates digital IC families (DTL, TTL, ECL, MOS, CMOS). Combinational circuits: arithmetic circuits, code converters, multiplexers and decoders. Sequential circuits: latches and flip-flops, counters and shift-registers. Sample and hold circuits, ADCs, DACs. Semiconductor memories. Microprocessor 8086: architecture, programming, memory and I/O interfacing.

Signals and Systems

Definitions and properties of Laplace transform, continuous-time and discrete-time Fourier series, continuous-time and discrete-time Fourier Transform, z-transform. Sampling theorems. Linear Time-Invariant (LTI) Systems: definitions and properties; causality, stability, impulse response, convolution, poles and zeros frequency response, group delay, phase delay. Signal transmission through LTI systems. Random signals and noise: probability, random variables, probability density function, autocorrelation, power spectral density, function analogy between vectors & functions.

Control Systems

Basic control system components; block diagrammatic description, reduction of block diagrams — Mason's rule. Open loop and closed loop (negative unity feedback) systems and stability analysis of these systems. Signal flow graphs and their use in determining transfer functions of systems; transient and steady state analysis of LTI control systems and frequency response. Analysis of steady-state disturbance rejection and noise sensitivity.

Tools and techniques for LTI control system analysis and design: root loci, Routh-Hurwitz stability criterion, Bode and Nyquist plots. Control system compensators: elements of lead and lag compensation, elements of Proportional-Integral-Derivative controller (PID). Discretization of continuous time systems using Zero-order hold (ZOH) and ADCs for digital controller implementation. Limitations of digital controllers: aliasing. State variable representation and solution of state equation of LTI control systems. Linearization of Nonlinear dynamical systems with state-space realizations in both frequency and time domains. Fundamental concepts of controllability and observability for MIMO LTI systems. State space realizations: observable and controllable canonical form. Ackermann's formula for state-feedback pole placement. Design of full order and reduced order estimators.

Communications

Analog communication systems: amplitude and angle modulation and demodulation systems, spectral analysis of these operations, superheterodyne noise conditions.

Digital communication systems: pulse code modulation (PCM), Differential Pulse Code Modulation (DPCM), Delta modulation (DM), digital modulation schemes-amplitude, phase and frequency shift keying schemes (ASK, PSK, FSK), matched filter receivers, bandwidth consideration and probability of error calculations for these schemes, GSM, TDMA.

Education and Training

Electronics engineers typically possess an academic degree with a major in electronic engineering. The length of study for such a degree is usually three or four years and the completed degree may be designated as a Bachelor of Engineering, Bachelor of Science, Bachelor of Applied Science, or Bachelor of Technology depending upon the university. Many UK universities also offer Master of Engineering (MEng) degrees at undergraduate level.

The degree generally includes units covering physics, chemistry, mathematics, project management and specific topics in electrical engineering. Initially such topics cover most, if not all, of the subfields of electronic engineering. Students then choose to specialize in one or more subfields towards the end of the degree.

Some electronics engineers also choose to pursue a postgraduate degree such as a Master of Science (MSc), Doctor of Philosophy in Engineering (PhD), or an Engineering Doctorate (EngD). The master's degree is being introduced in some European and American Universities as a first degree and the differentiation of an engineer with graduate and postgraduate studies is often difficult. In these cases, experience is taken into account. The master's degree may consist of either research, coursework or a mixture of the two. The Doctor of Philosophy consists of a significant research component and is often viewed as the entry point to academia.

In most countries, a bachelor's degree in engineering represents the first step towards certification and the degree program itself is certified by a professional body. After completing a certified degree program the engineer must satisfy a range of requirements (including work experience requirements) before being certified. Once certified the engineer is designated the title of Professional Engineer (in the United States, Canada and South Africa), Chartered Engineer or Incorporated Engineer (in the United Kingdom, Ireland, India and Zimbabwe), Chartered Professional Engineer (in Australia and New Zealand) or European Engineer (in much of the European Union).

Fundamental to the discipline are the sciences of physics and mathematics as these help to obtain both a qualitative and quantitative description of how such systems will work. Today most engineering work involves the use of computers and it is commonplace to use computer-aided design and simulation software programs when designing electronic systems. Although most electronic engineers will understand basic circuit theory, the theories employed by engineers generally depend upon the work they do. For example, quantum mechanics and solid state physics might be relevant to an engineer working on VLSI but are largely irrelevant to engineers working with macroscopic electrical systems.

Professional Bodies

Professional bodies of note for electrical engineers include the Institute of Electrical and Electronics Engineers (IEEE) and the Institution of Electrical Engineers (IEE) (now renamed the Institution of Engineering and Technology or IET). Member of the Institution of Engineering and Technology (MIET) is recognised in Europe as Electrical and computer (technology) engineer. The IEEE claims to produce 30 percent of the world's literature in electrical/electronic engineering, has over 430,000 members, and holds more than 450 IEEE sponsored or cosponsored conferences worldwide each year.

Project Engineering

For most engineers not involved at the cutting edge of system design and development, technical work accounts for only a fraction of the work they do. A lot of time is also spent on tasks such as discussing proposals with clients, preparing budgets and determining project schedules. Many senior engineers manage a team of technicians or other engineers and for this reason project management skills are important. Most engineering projects involve some form of documentation and strong written communication skills are therefore very important.

The workplaces of electronics engineers are just as varied as the types of work they do. Electronics engineers may be found in the pristine laboratory environment of a fabrication plant, the offices of a consulting firm or in a research laboratory. During their working life, electronics engineers may find themselves supervising a wide range of individuals including scientists, electricians, computer programmers and other engineers.

Obsolescence of technical skills is a serious concern for electronics engineers. Membership and participation in technical societies, regular reviews of periodicals in the field and a habit of continued learning are therefore essential to maintaining proficiency. And these are mostly used in the field of consumer electronics products.

Telecommunications Engineering

Telecommunications engineer working to maintain London's phone service during World War 2, January 1942 while all the power was off so he needed to find a power supply

'*Telecommunications engineering, or telecoms engineering*', is an engineering discipline that brings together all electrical engineering disciplines including computer engineering with systems engineering to enhance telecommunication systems. The work ranges from basic circuit design to strategic mass developments. A telecommunication engineer is responsible for designing and overseeing the installation of telecommunications equipment and facilities, such as complex electronic switching systems, copper wire telephone facilities,fiber optics cabling, IP data systems, Terrestrial radio link systems for conventional communications and process information. Telecommunication engineering also overlaps heavily with broadcast engineering.

Telecommunication is a diverse field of engineering which is connected to electronics, civil, structural, and electrical engineering. Ultimately, telecom engineers are responsible for providing the method for customers to have telephone and high-speed data services. It helps people who are closely working in political and social fields, as well accounting and project management.

Telecom engineers use a variety of equipment and transport media available from a multitude of manufacturers to design the telecom network infrastructure. The most common media used by wired telecommunications companies today are copper wires, coaxial cable, and fiber optics. Telecommunications engineers use their technical expertise to also provide a range of services and engineering solutions revolving around wireless mode of communication and other information transfer, such as wireless telephony services, radio and satellite communications, internet and broadband technologies.

Most of a telecom engineer's work is carried out on a project basis with tight deadlines and well-defined milestones for the delivery of project objectives. Telecommunication engineers are involved across all aspects of service delivery, from carrying out feasibility exercises and determining connectivity to preparing detailed, technical and operational documentation. This often leads to creative solutions to problems that often would have been designed differently without the budget constraints dictated by modern society. In the earlier days of the telecom industry, massive amounts of cable were placed that were never used or have been replaced by modern technology such as fiber optic cable and digital multiplexing techniques.

Telecom engineers are also responsible for overseeing the companies' records of equipment and facility assets. Their work directly impacts assigning appropriate accounting codes for taxes and maintenance purposes, budgeting and overseeing projects.

History

Telecommunication systems are generally designed by telecommunication engineers which sprang from technological improvements in the telegraph industry in the late 19th century and the radio and the telephone industries in the early 20th century. Today, telecommunication is widespread and devices that assist the process, such as the television, radio and telephone, are common in many parts of the world. There are also many networks that connect these devices, including computer networks, public switched telephone network (PSTN), radio networks, and television networks. Computer communication across the Internet is one of many examples of telecommunication. Telecommunication plays a vital role in the part of world economy and the telecommunication industry's revenue has been placed at just under 3% of the gross world product.

Telegraph and Telephone

Alexander Graham Bell's big box telephone, 1876, one of the first commercially available telephones - National Museum of American History

Samuel Morse independently developed a version of the electrical telegraph that he unsuccessfully demonstrated on 2 September 1837. Soon after he was joined by Alfred Vail who developed the register — a telegraph terminal that integrated a logging device for recording messages to paper tape. This was demonstrated successfully over three miles (five kilometres) on 6 January 1838 and eventually over forty miles (sixty-four kilometres) between Washington, D.C. and Baltimore on 24 May 1844. The patented invention proved lucrative and by 1851 telegraph lines in the United States spanned over 20,000 miles (32,000 kilometres).

The first successful transatlantic telegraph cable was completed on 27 July 1866, allowing transatlantic telecommunication for the first time. Earlier transatlantic cables installed in 1857 and 1858 only operated for a few days or weeks before they failed. The international use of the telegraph has sometimes been dubbed the "Victorian Internet".

The first commercial telephone services were set up in 1878 and 1879 on both sides of the Atlantic in the cities of New Haven and London. Alexander Graham Bell held the master patent for the telephone that was needed for such services in both countries. The technology grew quickly from this point, with inter-city lines being built and telephone exchanges in every major city of the United States by the mid-1880s. Despite this, transatlantic voice communication remained impossible for customers until January 7, 1927 when a connection was established using radio. However no cable connection existed until TAT-1 was inaugurated on September 25, 1956 providing 36 telephone circuits.

In 1880, Bell and co-inventor Charles Sumner Tainter conducted the world's first wireless telephone call via modulated lightbeams projected by photophones. The scientific principles of their invention would not be utilized for several decades, when they were first deployed in military and fiber-optic communications.

Radio and Television

Marconi crystal radio receiver

Over several years starting in 1894 the Italian inventor Guglielmo Marconi built the first complete, commercially successful wireless telegraphy system based on airborne electromagnetic waves (radio transmission). In December 1901, he would go on to established wireless communication between Britain and Newfoundland, earning him the Nobel Prize in physics in 1909 (which he shared with Karl Braun). In 1900 Reginald Fessenden was able to wirelessly transmit a human voice. On March 25, 1925, Scottish inventor John Logie Baird publicly demonstrated the transmission of moving silhouette pictures at the London department store Selfridges. In October 1925, Baird was successful in obtaining moving pictures with halftone shades, which were by most accounts the first true television pictures. This led to a public demonstration of the improved device on 26 January 1926 again at Selfridges. Baird's first devices relied upon the Nipkow disk and thus became known as the mechanical television. It formed the basis of semi-experimental broadcasts done by the British Broadcasting Corporation beginning September 30, 1929.

Satellite

The first U.S. satellite to relay communications was Project SCORE in 1958, which used a tape recorder to store and forward voice messages. It was used to send a Christmas greeting to the world from U.S. President Dwight D. Eisenhower. In 1960 NASA launched an Echo satellite; the 100-foot (30 m) aluminized PET film balloon served as a passive reflector for radio communications. Courier 1B, built by Philco, also launched in 1960, was the world's first active repeater satellite. Satellites these days are used for many applications such as uses in GPS, television, internet and telephone uses.

Telstar was the first active, direct relay commercial communications satellite. Belonging to AT&T as part of a multi-national agreement between AT&T, Bell Telephone Laboratories, NASA, the British General Post Office, and the French National PTT (Post Office) to develop satellite communications, it was launched by NASA from Cape Canaveral on July 10, 1962, the first privately sponsored space launch. Relay 1 was launched on December 13, 1962, and became the first satellite to broadcast across the Pacific on November 22, 1963.

The first and historically most important application for communication satellites was in intercontinental long distance telephony. The fixed Public Switched Telephone Network relays telephone calls from land line telephones to an earth station, where they are then transmitted a receiving satellite dish via a geostationary satellite in Earth orbit. Improvements in submarine communications cables, through the use of fiber-optics, caused some decline in the use of satellites for fixed telephony in the late 20th century, but they still exclusively service remote islands such as Ascension Island, Saint Helena, Diego Garcia, and Easter Island, where no submarine cables are in service. There are also some continents and some regions of countries where landline telecommunications are rare to nonexistent, for example Antarctica, plus large regions of Australia, South America, Africa, Northern Canada, China, Russia and Greenland.

After commercial long distance telephone service was established via communication satellites, a host of other commercial telecommunications were also adapted to similar satellites starting in 1979, including mobile satellite phones, satellite radio, satellite television and satellite Internet access. The earliest adaption for most such services occurred in the 1990s as the pricing for commercial satellite transponder channels continued to drop significantly.

Computer Networks and the Internet

On 11 September 1940, George Stibitz was able to transmit problems using teleprinter to his Complex Number Calculator in New York and receive the computed results back at Dartmouth College in New Hampshire. This configuration of a centralized computer or mainframe computer with remote "dumb terminals" remained popular throughout the 1950s and into the 1960s. However, it was not until the 1960s that researchers started to investigate packet switching — a technology that allows chunks of data to be sent between different computers without first passing through a centralized mainframe. A four-node network emerged on 5 December 1969. This network soon became the ARPANET, which by 1981 would consist of 213 nodes.

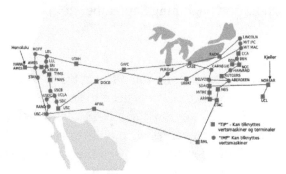

Symbolic representation of the Arpanet as of September 1974

ARPANET's development centered around the Request for Comment process and on 7 April 1969, RFC 1 was published. This process is important because ARPANET would eventually merge with other networks to form the Internet, and many of the communication protocols that the Internet relies upon today were specified through the Request for Comment process. In September 1981, RFC 791 introduced the Internet Protocol version 4 (IPv4) and RFC 793 introduced the Transmission Control Protocol (TCP) — thus creating the TCP/IP protocol that much of the Internet relies upon today.

Optical Fiber

Optical fiber can be used as a medium for telecommunication and computer networking because it is flexible and can be bundled as cables. It is especially advantageous for long-distance communications, because light propagates through the fiber with little attenuation compared to electrical cables. This allows long distances to be spanned with few repeaters.

In 1966 Charles K. Kao and George Hockham proposed optical fibers at STC Laboratories (STL) at Harlow, England, when they showed that the losses of 1000 dB/km in existing glass (compared to 5-10 dB/km in coaxial cable) was due to contaminants, which could potentially be removed.

Optical fiber was successfully developed in 1970 by Corning Glass Works, with attenuation low enough for communication purposes (about 20dB/km), and at the same time GaAs (Gallium arsenide) semiconductor lasers were developed that were compact and therefore suitable for transmitting light through fiber optic cables for long distances.

After a period of research starting from 1975, the first commercial fiber-optic communications system was developed, which operated at a wavelength around 0.8 µm and used GaAs semiconductor lasers. This first-generation system operated at a bit rate of 45 Mbps with repeater spacing of up to 10 km. Soon on 22 April 1977, General Telephone and Electronics sent the first live telephone traffic through fiber optics at a 6 Mbit/s throughput in Long Beach, California.

The first wide area network fibre optic cable system in the world seems to have been installed by Rediffusion in Hastings, East Sussex, UK in 1978. The cables were placed in ducting throughout the town, and had over 1000 subscribers. They were used at that time for the transmission of television channels,not available because of local reception problems.

The first transatlantic telephone cable to use optical fiber was TAT-8, based on Desurvire optimized laser amplification technology. It went into operation in 1988.

In the late 1990s through 2000, industry promoters, and research companies such as KMI, and RHK predicted massive increases in demand for communications bandwidth due to increased use of the Internet, and commercialization of various bandwidth-intensive consumer services, such as video on demand. Internet protocol data traffic was increasing exponentially, at a faster rate than integrated circuit complexity had increased under Moore's Law.

Concepts

Radio Transmitter room

Basic Elements of a Telecommunication System

Transmitter

Transmitter (information source) that takes information and converts it to a signal for transmission. In electronics and telecommunications a transmitter or radio transmitter is an electronic device which, with the aid of an antenna, produces radio waves. In addition to their use in broadcasting, transmitters are necessary component parts of many electronic devices that communicate by radio, such as cell phones,

Copper wires

Transmission Medium

Transmission medium over which the signal is transmitted. For example, the transmission medium for sounds is usually air, but solids and liquids may also act as transmission media for sound. Many transmission media are used as communications channel. One of the most common physical medias used in networking is copper wire. Copper wire to carry signals to long distances using relatively low amounts of power.Another example of a physical medium is optical fiber, which has emerged as the most commonly used transmission medium for long-distance communications. Optical fiber is a thin strand of glass that guides light along its length.

The absence of a material medium in vacuum may also constitute a transmission medium for electromagnetic waves such as light and radio waves.

Receiver

Receiver (information sink) that receives and converts the signal back into required information. In radio communications, a radio receiver is an electronic device that receives radio waves and converts the information carried by them to a usable form. It is used with an antenna. The information produced by the receiver may be in the form of sound (an audio signal), images (a video signal) or digital data.

Wireless communication tower, cell site

Wired Communication

Wired communications make use of underground communications cables (less often, overhead lines), electronic signal amplifiers (repeaters) inserted into connecting cables at specified points, and terminal apparatus of various types, depending on the type of wired communications used.

Wireless Communication

Wireless communication involves the transmission of information over a distance without help of wires, cables or any other forms of electrical conductors. Wireless operations permit services, such as long-range communications, that are impossible or impractical to implement with the use of wires. The term is commonly used in the telecommunications industry to refer to telecommunications systems (e.g. radio transmitters and receivers, remote controls etc.) which use some form of energy (e.g. radio waves, acoustic energy, etc.) to transfer information without the use of wires. Information is transferred in this manner over both short and long distances.

Roles

Telecom Equipment Engineer

A telecom equipment engineer is an electronics engineer that designs equipment such as routers, switches, multiplexers, and other specialized computer/electronics equipment designed to be used in the telecommunication network infrastructure.

Network Engineer

A network engineer is a computer engineer who is in charge of designing, deploying and maintaining computer networks. In addition, they oversee network operations from a network operations center, designs backbone infrastructure, or supervises interconnections in a data center.

Central-office Engineer

Typical Northern Telecom DMS100 Telephone Central Office Installation

A central-office engineer is responsible for designing and overseeing the implementation of telecommunications equipment in a central office (CO for short), also referred to as a wire center or telephone exchange A CO engineer is responsible for integrating new technology into the existing network, assigning the equipment's location in the wire center, and providing power, clocking (for digital equipment), and alarm monitoring facilities for the new equipment. The CO engineer is also responsible for providing more power, clocking, and alarm monitoring facilities if there are currently not enough available to support the new equipment being installed. Finally, the CO engineer is responsible for designing how the massive amounts of cable will be distributed to various equipment and wiring frames throughout the wire center and overseeing the installation and turn up of all new equipment.

Subroles

As structural engineers, CO engineers are responsible for the structural design and placement of racking and bays for the equipment to be installed in as well as for the plant to be placed on.

As electrical engineers, CO engineers are responsible for the resistance, capacitance, and inductance (RCL) design of all new plant to ensure telephone service is clear and crisp and data service is clean as well as reliable. Attenuation or gradual loss in intensity and loop loss calculations are required to determine cable length and size required to provide the service called for. In addition, power requirements have to be calculated and provided to power any electronic equipment being placed in the wire center.

Overall, CO engineers have seen new challenges emerging in the CO environment. With the advent of Data Centers, Internet Protocol (IP) facilities, cellular radio sites, and other emerging-technology equipment environments within telecommunication networks, it is important that a consistent set of established practices or requirements be implemented.

Installation suppliers or their sub-contractors are expected to provide requirements with their products, features, or services. These services might be associated with the installation of new or expanded equipment, as well as the removal of existing equipment.

Several other factors must be considered such as:

- Regulations and safety in installation

- Removal of hazardous material

- Commonly used tools to perform installation and removal of equipment

Outside-plant Engineer

Engineers working on a cross-connect box, also known as a serving area interface.

Outside plant (OSP) engineers are also often called field engineers because they frequently spend much time in the field taking notes about the civil environment, aerial, above ground, and below ground. OSP engineers are responsible for taking plant (copper, fiber, etc.) from a wire center to a distribution point or destination point directly. If a distribution point design is used, then a cross-connect box is placed in a strategic location to feed a determined distribution area.

The cross-connect box, also known as a serving area interface, is then installed to allow connections to be made more easily from the wire center to the destination point and ties up fewer facilities by not having dedication facilities from the wire center to every destination point. The plant is then taken directly to its destination point or to another small closure called a terminal, where access can also be gained to the plant if necessary. These access points are preferred as they allow faster repair times for customers and save telephone operating companies large amounts of money.

The plant facilities can be delivered via underground facilities, either direct buried or through conduit or in some cases laid under water, via aerial facilities such as telephone or power poles, or via microwave radio signals for long distances where either of the other two methods is too costly.

Subroles

Engineer (OSP) climbing the telephone pole.

As structural engineers, OSP engineers are responsible for the structural design and placement of cellular towers and telephone poles as well as calculating pole capabilities of existing telephone or power poles onto which new plant is being added. Structural calculations are required when boring under heavy traffic areas such as highways or when attaching to other structures such as bridges. Shoring also has to be taken into consideration for larger trenches or pits. Conduit structures often include encasements of slurry that needs to be designed to support the structure and withstand the environment around it (soil type, high traffic areas, etc.).

As electrical engineers, OSP engineers are responsible for the resistance, capacitance, and inductance (RCL) design of all new plant to ensure telephone service is clear and crisp and data service is clean as well as reliable. Attenuation or gradual loss in intensity and loop loss calculations are required to determine cable length and size required to provide the service called for. In addition power requirements have to be calculated and provided to power any electronic equipment being placed in the field. Ground potential has to be taken into consideration when placing equipment, facilities, and plant in the field to account for lightning strikes, high voltage intercept from improperly grounded or broken power company facilities, and from various sources of electromagnetic interference.

As civil engineers, OSP engineers are responsible for drafting plans, either by hand or using Computer-aided design (CAD) software, for how telecom plant facilities will be placed. Often when working with municipalities trenching or boring permits are required and drawings must be made for these. Often these drawings include about 70% or so of the detailed information required to pave a road or add a turn lane to an existing street. Structural calculations are required when boring under heavy traffic areas such as highways or when attaching to other structures such as bridges. As civil engineers, telecom engineers provide the modern communications backbone for all technological communications distributed throughout civilizations today.

Unique to telecom engineering is the use of air-core cable which requires an extensive network of air handling equipment such as compressors, manifolds, regulators and hundreds of miles of air pipe per system that connects to pressurized splice cases all designed to pressurize this special form of copper cable to keep moisture out and provide a clean signal to the customer.

As political and social ambassador, the OSP engineer is a telephone operating company's face and voice to the local authorities and other utilities. OSP engineers often meet with municipalities, construction companies and other utility companies to address their concerns and educate them about how the telephone utility works and operates. Additionally, the OSP engineer has to secure real estate to place outside facilities on, such as an easement to place a cross-connect box on.

Computer Engineering

The motherboard used in a HD DVD player, the result of computer engineering efforts.

Computer engineering is a discipline that integrates several fields of electrical engineering and computer science required to develop computer hardware and software. Computer engineers usually have training in electronic engineering (or electrical engineering), software design, and hardware-software integration instead of only software engineering or electronic engineering. Computer engineers are involved in many hardware and software aspects of computing, from the design of individual microcontrollers, microprocessors, personal computers, and supercomputers, to circuit design. This field of engineering not only focuses on how computer systems themselves work, but also how they integrate into the larger picture.

Usual tasks involving computer engineers include writing software and firmware for embedded microcontrollers, designing VLSI chips, designing analog sensors, designing mixed signal circuit boards, and designing operating systems. Computer engineers are also suited for robotics research, which relies heavily on using digital systems to control and monitor electrical systems like motors, communications, and sensors.

In many institutions, computer engineering students are allowed to choose areas of in-depth study in their junior and senior year, because the full breadth of knowledge used in the design and application of computers is beyond the scope of an undergraduate degree. Other institutions may require engineering students to complete one year of General Engineering before declaring computer engineering as their primary focus.

History

The first computer engineering degree program in the United States was established at Case Western Reserve University in 1972. As of 2015, there were 238 ABET-accredited computer engineering programs in the US. In Europe, accreditation of computer engineering schools is done by a variety of agencies part of the EQANIE network. Due to increasing job requirements for engineers who can concurrently design hardware, software, firmware, and manage all forms of computer systems used in industry, some tertiary institutions around the world offer a bachelor's degree generally called computer engineering. Both computer engineering and electronic engineering programs include analog and digital circuit design in their curriculum. As with most engineering disciplines, having a sound knowledge of mathematics and science is necessary for computer engineers.

Work

There are two major specialties in computer engineering: software and hardware.

Computer Software Engineering

Computer software engineers develop, design, and test software. Some software engineers design, construct, and maintain computer programs for companies. Some set up networks such as "intranets" for companies. Others make or install new software or upgrade computer systems. Computer software engineers can also work in application design. This involves designing or coding new programs and applications to meet the needs of a business or individual. Computer software engineers can also work as freelancers and sell their software products/applications to an enterprise/individual.

Computer Hardware Engineering

Most computer hardware engineers research, develop, design, and test various computer equipment. This can range from circuit boards and microprocessors to routers. Some update existing computer equipment to be more efficient and work with newer software. Most computer hardware engineers work in research laboratories and high-tech manufacturing firms. Some also work for the federal government. According to BLS, 95% of computer hardware engineers work in metropolitan areas. They generally work full-time. Approximately 33% of their work requires more than 40 hours a week. The median salary for employed qualified computer hardware engineers (2012) was $100,920 per year or $48.52 per hour. Computer hardware engineers held 83,300 jobs in 2012 in the USA.

Specialty Areas

There are many specialty areas in the field of computer engineering.

Coding, Cryptography, and Information Protection

Computer engineers work in coding, cryptography, and information protection to develop new methods for protecting various information, such as digital images and music, fragmentation, copyright infringement and other forms of tampering. Examples include work on wireless communications, multi-antenna systems, optical transmission, and digital watermarking.

Communications and Wireless Networks

Those focusing on communications and wireless networks, work advancements in telecommunications systems and networks (especially wireless networks), modulation and error-control coding, and information theory. High-speed network design, interference suppression and modulation, design and analysis of fault-tolerant system, and storage and transmission schemes are all a part of this specialty.

Compilers and Operating Systems

This specialty focuses on compilers and operating systems design and development. Engineers in this field develop new operating system architecture, program analysis techniques, and new techniques to assure quality. Examples of work in this field includes post-link-time code transformation algorithm development and new operating system development.

Computational Science and Engineering

Computational Science and Engineering is a relatively new discipline. According to the Sloan Career Cornerstone Center, individuals working in this area, "computational methods are applied to formulate and solve complex mathematical problems in engineering and the physical and the social sciences. Examples include aircraft design, the plasma processing of nanometer features on semiconductor wafers, VLSI circuit design, radar detection systems, ion transport through biological channels, and much more".

Computer Networks, Mobile Computing, and Distributed Systems

In this specialty, engineers build integrated environments for computing, communications, and information access. Examples include shared-channel wireless networks, adaptive resource management in various systems, and improving the quality of service in mobile and ATM environments. Some other examples include work on wireless network systems and fast Ethernet cluster wired systems.

Computer Systems: Architecture, Parallel Processing, and Dependability

Engineers working in computer systems work on research projects that allow for reliable, secure, and high-performance computer systems. Projects such as designing processors for multi-threading and parallel processing are included in this field. Other examples of work in this field include development of new theories, algorithms, and other tools that add performance to computer systems.

Computer Vision and Robotics

In this specialty, computer engineers focus on developing visual sensing technology to sense an environment, representation of an environment, and manipulation of the environment. The gathered three-dimensional information is then implemented to perform a variety of tasks. These include, improved human modeling, image communication, and human-computer interfaces, as well as devices such as special-purpose cameras with versatile vision sensors.

Embedded Systems

Examples of devices that use embedded systems.

Individuals working in this area design technology for enhancing the speed, reliability, and performance of systems. Embedded systems are found in many devices from a small FM radio to the space shuttle. According to the Sloan Cornerstone Career Center, ongoing developments in embedded systems include "automated vehicles and equipment to conduct search and rescue, automated transportation systems, and human-robot coordination to repair equipment in space."

Integrated Circuits, VLSI Design, Testing and CAD

This specialty of computer engineering requires adequate knowledge of electronics and electrical systems. Engineers working in this area work on enhancing the speed, reliability, and energy efficiency of next-generation very-large-scale integrated (VLSI) circuits and microsystems. An example of this specialty is work done on reducing the power consumption of VLSI algorithms and architecture.

Signal, Image and Speech Processing

Computer engineers in this area develop improvements in human–computer interaction, including speech recognition and synthesis, medical and scientific imaging, or communications systems. Other work in this area includes computer vision development such as recognition of human facial features.

Education

Most entry-level computer engineering jobs require at least a bachelor's degree in computer engineering. Sometimes a degree in electronic engineering is accepted, due to the similarity of the two fields. Because hardware engineers commonly work with computer software systems, a background in computer programming usually is needed. According to BLS, "a computer engineering major is similar to electrical engineering but with some computer science courses added to the curriculum". Some large firms or specialized jobs require a master's degree. It is also important for computer engineers to keep up with rapid advances in technology. Therefore, many continue learning throughout their careers.

Job Outlook in the United States

Computer Software Engineering

According to the U.S. Bureau of Labor Statistics (BLS), "computer applications software engineers and computer systems software engineers are projected to be among the faster than average growing occupations" from 2014-24, with a projected growth rate of 17%. This is down from the 2012 to 2022 BLS estimate of 22% for software developers. And, further down from the 30% 2010 to 2020 BLS estimate. In addition, growing concerns over cyber security add up to put computer software engineering high above the average rate of increase for all fields. However, some of the work will be outsourced in foreign countries. Due to this, job growth will not be as fast as during the last decade, as jobs that would have gone to computer software engineers in the United States would instead go to computer software engineers in countries such as India. In addition the BLS Job Outlook for Computer Programmers, 2014-24 has an -8% (a decline in their words) for those who program computers (i.e. embedded systems) who are not computer application developers.

Computer Hardware Engineering

According to the BLS, Job Outlook employment for computer hardware engineers, 2014-24 is 3% ("Slower than average" in their own words when compared to other occupations)" and is down from 7% for 2012 to 2022 BLS estimate and is further down from 9% in the BLS 2010 to 2020 estimate." Today, computer hardware is somehow equal to Electronic and Computer Engineering (ECE) and has divided to many subcategories, the most significant of them is Embedded system design.

Similar Occupations and Fields

- Computer programming
- Electrical engineering
- Software development
- Systems analyst

Electromechanics

In engineering, electromechanics combines electrical and mechanical processes and procedures drawn from electrical engineering and mechanical engineering. Electrical engineering in this context also encompasses electronics engineering.

Devices which carry out electrical operations by using moving parts are known as electromechanical. Strictly speaking, a manually operated switch is an electromechanical component, but the term is usually understood to refer to devices which involve an electrical signal to create mechanical movement, or mechanical movement to create an electric signal. Often involving electromagnetic principles such as in relays, which allow a voltage or current to control other, usually isolated circuit voltage or current by mechanically switching sets of contacts, and solenoids, by which a voltage can actuate a moving linkage as in solenoid valves. Piezoelectric devices are electromechanical, but do not use electromagnetic principles. Piezoelectric devices can create sound or vibration from an electrical signal or create an electrical signal from sound or mechanical vibration.

Before the development of modern electronics, electromechanical devices were widely used in complicated systems subsystems, including electric typewriters, teleprinters, very early television systems, and the very early electromechanical digital computers.

History of Electromechanics

Relays originated with telegraphy as electromechanical devices used to regenerate telegraph signals. In 1885, Michael Pupin at Columbia University taught mathematical physics and electromechanics until 1931.

The Strowger switch, the Panel switch, and similar ones were widely used in early automated telephone exchanges. Crossbar switches were first widely installed in the middle 20th century in Sweden, the United States, Canada, and Great Britain, and these quickly spread to the rest of the world - especially to Japan. The electromechanical television systems of the late 19th century were less successful.

Electric typewriters developed, up to the 1980s, as "power-assisted typewriters". They contained a single electrical component, the motor. Where the keystroke had previously moved a typebar directly, now it engaged mechanical linkages that directed mechanical power from the motor into the typebar. This was also true of the later IBM Selectric. At Bell Labs, in the 1940s, the Bell Model V computer was developed. It was an electromechanical relay-based device; cycles took seconds. In 1968 electromechanical systems were still under serious consideration for an aircraft flight control computer, until a device based on large scale integration electronics was adopted in the Central Air Data Computer.

Modern Practice

Beginning in the last third of the century, much equipment which for most of the 20th century would have used electromechanical devices for control, has come to use less expensive and more reliable integrated microcontroller circuits containing ultimately a few million transistors, and a program to carry out the same task through logic, with electromechanical components only where moving parts, such as mechanical electric actuators, are a requirement. Such chips have replaced most electromechanical devices, because any point in a system which must rely on mechanical movement for proper operation will have mechanical wear and eventually fail. Properly designed electronic circuits without moving parts will continue to operate properly almost indefinitely and are used in most simple feedback control systems, and appear in huge numbers in everything from traffic lights to washing machines.

As of 2010, approximately 16,400 people work as electro-mechanical technicians in the US, about 1 out of every 9000 workers. Their median annual wage is about 50% more than the median annual wage over all occupations.

Radio-frequency Engineering

Radio-frequency engineering is a subset of electrical engineering that deals with devices that are designed to operate in the radio frequency (RF) spectrum. These devices operate within the range of about 3 kHz up to 300 GHz.

Radio-frequency engineering is incorporated into almost everything that transmits or receives a radio wave, which includes, but is not limited to, mobile phones, radios, Wi-Fi, and two-way radios.

Radio-frequency engineering is a highly specialized field falling typically in one of two areas:

- providing or controlling coverage with some kind of antenna/transmission system

- generating or receiving signals to or from that transmission system to other communications electronics or controls.

To produce quality results, an in-depth knowledge of mathematics, physics, general electronics theory as well as specialized training in areas such as wave propagation, impedance transformations, filters, microstrip circuit board design, etc. may be required. Because of the many ways RF is conducted both through typical conductors as well as through space, an initial design of an RF circuit usually bears very little resemblance to the final optimized physical circuit. Revisions to the design are often required to achieve intended results.

Radio Electronics

Radio electronics is concerned with electronic circuits which receive or transmit radio signals.

Typically, such circuits must operate at radio frequency and power levels, which imposes special constraints on their design. These constraints increase in their importance with higher frequencies. At microwave frequencies, the reactance of signal traces becomes a crucial part of the physical layout of the circuit.

List of radio electronics topics:

- RF oscillators: PLL, Voltage-controlled oscillator

- Transmitters, transmission lines, RF connectors

- Antennas, antenna theory, list of antenna terms

- Receivers, tuners

- Amplifiers

- Modulators, demodulators, detectors

- RF filters

- RF shielding, Ground plane

- PCB layout guidelines

- DSSS, noise power
- Digital radio

Duties

Radio-frequency engineers are specialists in their respective field and can take on many different roles, such as design, installation, and maintenance. Radio-frequency engineers require many years of extensive experience in the area of study. This type of engineer has experience with transmission systems, device design, and placement of antennas for optimum performance. A radio-frequency engineer at a broadcast facility is responsible for maintenance of the stations high-power broadcast transmitters and associated systems. This includes transmitter site emergency power, remote control, main transmission line and antenna adjustments, microwave radio relay STL/TSL links, and more.

In addition, a radio-frequency design engineer must be able to understand electronic hardware design, circuit board material, antenna radiation, and the effect of interfering frequencies that prevent optimum performance within the piece of equipment being developed.

Early Radio-frequency Engineers

Many notable individuals have contributed to the advancement of Radio-frequency engineering theory and design, including the following:

- Heinrich Hertz, demonstrated the existence of radio waves and developed the unit of measure to describe frequency of a wave.

- Nikola Tesla, known for his high-voltage, high-frequency power experiments in New York and Colorado Springs. Tesla's primary interest was wireless power transmission through a medium (primarily the Earth) with demonstrations in 1893 in St. Louis, Missouri, at the Franklin Institute in Philadelphia, Pennsylvania, and at the National Electric Light Association but saw communication as a side aspect.

- Guglielmo Marconi, who developed the first successful commercial wireless telegraphy system based on air-born radio frequency waves (called Herzian waves at the time) and transmitted the first radio signal across the Atlantic Ocean.

- Phillip H. Smith, who developed a graphical method of calculating impedances, admittances, reflection coefficients and scattering parameters.

References

- Anthony J. Pansini Electrical Distribution Engineering, p. xiv, The Fairmont Press Inc., 2006 ISBN 978-0-88173-546-8

- Electronics: Converters, Applications, and Design. United States of America: John Wiley & Sons, Inc. ISBN 0-471-22693-9.

- Charles A. Harper High Performance Printed Circuit Boards, pp. xiii–xiv, McGraw-Hill Professional, 2000 ISBN 978-0-07-026713-8

- Jimmie J. Cathey Schaum's Outline of Theory and Problems of Electronic Devices and Circuits, McGraw Hill,

2002 ISBN 978-0-07-136270-2

- Anant Agarwal/Jeffrey H. Lang Foundations of Analog and Digital Electronic Circuits, Morgan Kaufmann, 2005 ISBN 978-1-55860-735-4

- Hwei Piao Hsu Schaum's Outline of Theory and Problems of Signals and Systems, p. 1, McGraw–Hill Professional, 1995 ISBN 978-0-07-030641-7

- Gerald Luecke, Analog and Digital Circuits for Electronic Control System Applications, Newnes, 2005. ISBN 978-0-7506-7810-0.

- Joseph J. DiStefano, Allen R. Stubberud, and Ivan J. Williams, Schaum's Outline of Theory and Problems of Feedback and Control Systems, McGraw-Hill Professional, 1995. ISBN 978-0-07-017052-0.

- Hwei Pia Hsu, Schaum's Outline of Analog and Digital Communications, McGraw–Hill Professional, 2003. ISBN 978-0-07-140228-6.

- Homer L. Davidson, Troubleshooting and Repairing Consumer Electronics, p. 1, McGraw–Hill Professional, 2004. ISBN 978-0-07-142181-2.

Key Elements of Electrical Engineering

Electricity, electronics and electromagnetism are important topics related to the field of electrical engineering. Electronics controls electrical energy, it is concerned with components such as vacuum tubes and integrated circuits. The following chapter unfolds its key elements in a critical yet systematic manner.

Electricity

Lightning is one of the most dramatic effects of electricity.

Electricity is the set of physical phenomena associated with the presence and flow of electric charge. Electricity gives a wide variety of well-known effects, such as lightning, static electricity, electromagnetic induction and electric current. In addition, electricity permits the creation and reception of electromagnetic radiation such as radio waves.

In electricity, charges produce electromagnetic fields which act on other charges. Electricity occurs due to several types of physics:

- electric charge: a property of some subatomic particles, which determines their electromagnetic interactions. Electrically charged matter is influenced by, and produces, electromagnetic fields, electric charges can be positive or negative.

- electric field charges are surrounded by an electric field. The electric field produces a force on other charges. Changes in the electric field travel at the speed of light.

- electric potential: the capacity of an electric field to do work on an electric charge, typically measured in volts.

- electric current: a movement or flow of electrically charged particles, typically measured in amperes.

- electromagnets: Moving charges produce a magnetic field. Electric currents generate magnetic fields, and changing magnetic fields generate electric currents.

In electrical engineering, electricity is used for:

- electric power where electric current is used to energise equipment;

- electronics which deals with electrical circuits that involve active electrical components such as vacuum tubes, transistors, diodes and integrated circuits, and associated passive interconnection technologies.

Electrical phenomena have been studied since antiquity, though progress in theoretical understanding remained slow until the seventeenth and eighteenth centuries. Even then, practical applications for electricity were few, and it would not be until the late nineteenth century that engineers were able to put it to industrial and residential use. The rapid expansion in electrical technology at this time transformed industry and society. Electricity's extraordinary versatility means it can be put to an almost limitless set of applications which include transport, heating, lighting, communications, and computation. Electrical power is now the backbone of modern industrial society.

History

Thales, the earliest known researcher into electricity

Long before any knowledge of electricity existed, people were aware of shocks from electric fish. Ancient Egyptian texts dating from 2750 BCE referred to these fish as the "Thunderer of the Nile", and described them as the "protectors" of all other fish. Electric fish were again reported millennia later by ancient Greek, Roman and Arabic naturalists and physicians. Several ancient writers, such as Pliny the Elder and Scribonius Largus, attested to the numbing effect of electric shocks delivered by catfish and electric rays, and knew that such shocks could travel along conducting objects. Patients suffering from ailments such as gout or headache were directed to touch electric fish in the hope that the powerful jolt might cure them. Possibly the earliest and nearest approach to the discovery of the identity of lightning, and electricity from any other source, is to be attributed to the Arabs, who before the 15th century had the Arabic word for lightning (*raad*) applied to the electric ray.

Ancient cultures around the Mediterranean knew that certain objects, such as rods of amber, could be rubbed with cat's fur to attract light objects like feathers. Thales of Miletus made a series of observations on static electricity around 600 BCE, from which he believed that friction rendered amber magnetic, in contrast to minerals such as magnetite, which needed no rubbing. Thales was incorrect in believing the attraction was due to a magnetic effect, but later science would prove a link between magnetism and electricity. According to a controversial theory, the Parthians may have had knowledge of electroplating, based on the 1936 discovery of the Baghdad Battery, which resembles a galvanic cell, though it is uncertain whether the artifact was electrical in nature.

Benjamin Franklin conducted extensive research on electricity in the 18th century, as documented by Joseph Priestley (1767) *History and Present Status of Electricity*, with whom Franklin carried on extended correspondence.

Electricity would remain little more than an intellectual curiosity for millennia until 1600, when the English scientist William Gilbert made a careful study of electricity and magnetism, distinguishing the lodestone effect from static electricity produced by rubbing amber. He coined the New Latin word *electricus* to refer to the property of attracting small objects after being rubbed. This association gave rise to the English words "electric" and "electricity", which made their first appearance in print in Thomas Browne's *Pseudodoxia Epidemica* of 1646.

Further work was conducted by Otto von Guericke, Robert Boyle, Stephen Gray and C. F. du Fay. In the 18th century, Benjamin Franklin conducted extensive research in electricity, selling his possessions to fund his work. In June 1752 he is reputed to have attached a metal key to the bottom of a dampened kite string and flown the kite in a storm-threatened sky. A succession of sparks jumping from the key to the back of his hand showed that lightning was indeed electrical in nature. He also explained the apparently paradoxical behavior of the Leyden jar as a device for storing large amounts of electrical charge in terms of electricity consisting of both positive and negative charges.

Michael Faraday's discoveries formed the foundation of electric motor technology

In 1791, Luigi Galvani published his discovery of bioelectromagnetics, demonstrating that electricity was the medium by which neurons passed signals to the muscles. Alessandro Volta's battery, or voltaic pile, of 1800, made from alternating layers of zinc and copper, provided scientists with a more reliable source of electrical energy than the electrostatic machines previously used. The recognition of electromagnetism, the unity of electric and magnetic phenomena, is due to Hans Christian Ørsted and André-Marie Ampère in 1819-1820; Michael Faraday invented the electric motor in 1821, and Georg Ohm mathematically analysed the electrical circuit in 1827. Electricity and magnetism (and light) were definitively linked by James Clerk Maxwell, in particular in his "On Physical Lines of Force" in 1861 and 1862.

While the early 19th century had seen rapid progress in electrical science, the late 19th century would see the greatest progress in electrical engineering. Through such people as Alexander Graham Bell, Ottó Bláthy, Thomas Edison, Galileo Ferraris, Oliver Heaviside, Ányos Jedlik, William Thomson, 1st Baron Kelvin, Charles Algernon Parsons, Werner von Siemens, Joseph Swan, Nikola Tesla and George Westinghouse, electricity turned from a scientific curiosity into an essential tool for modern life, becoming a driving force of the Second Industrial Revolution.

In 1887, Heinrich Hertz discovered that electrodes illuminated with ultraviolet light create electric sparks more easily. In 1905 Albert Einstein published a paper that explained experimental data from the photoelectric effect as being the result of light energy being carried in discrete quantized packets, energising electrons. This discovery led to the quantum revolution. Einstein was awarded the Nobel Prize in Physics in 1921 for "his discovery of the law of the photoelectric effect". The photoelectric effect is also employed in photocells such as can be found in solar panels and this is frequently used to make electricity commercially.

The first solid-state device was the "cat's-whisker detector" first used in the 1900s in radio receivers. A whisker-like wire is placed lightly in contact with a solid crystal (such as a germanium crystal) in order to detect a radio signal by the contact junction effect. In a solid-state component, the current is confined to solid elements and compounds engineered specifically to switch and amplify it. Current flow can be understood in two forms: as negatively charged electrons, and as positively charged electron deficiencies called holes. These charges and holes are understood in terms of quantum physics. The building material is most often a crystalline semiconductor.

The solid-state device came into its own with the invention of the transistor in 1947. Common solid-state devices include transistors, microprocessor chips, and RAM. A specialized type of RAM called flash RAM is used in USB flash drives and more recently, solid-state drives to replace mechanically rotating magnetic disc hard disk drives. Solid state devices became prevalent in the 1950s and the 1960s, during the transition from vacuum tubes to semiconductor diodes, transistors, integrated circuit (IC) and the light-emitting diode (LED).

Concepts

Electric Charge

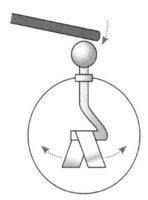

Charge on a gold-leaf electroscope causes the leaves to visibly repel each other

The presence of charge gives rise to an electrostatic force: charges exert a force on each other, an effect that was known, though not understood, in antiquity. A lightweight ball suspended from a string can be charged by touching it with a glass rod that has itself been charged by rubbing with a cloth. If a similar ball is charged by the same glass rod, it is found to repel the first: the charge acts to force the two balls apart. Two balls that are charged with a rubbed amber rod also repel each other. However, if one ball is charged by the glass rod, and the other by an amber rod, the two balls are found to attract each other. These phenomena were investigated in the late eighteenth century by Charles-Augustin de Coulomb, who deduced that charge manifests itself in two opposing forms. This discovery led to the well-known axiom: *like-charged objects repel and opposite-charged objects attract.*

The force acts on the charged particles themselves, hence charge has a tendency to spread itself as evenly as possible over a conducting surface. The magnitude of the electromagnetic force, whether attractive or repulsive, is given by Coulomb's law, which relates the force to the product of the charges and has an inverse-square relation to the distance between them. The electromagnetic force is very strong, second only in strength to the strong interaction, but unlike that force it operates over all distances. In comparison with the much weaker gravitational force, the electromagnetic force pushing two electrons apart is 10^{42} times that of the gravitational attraction pulling them together.

Study has shown that the origin of charge is from certain types of subatomic particles which have the property of electric charge. Electric charge gives rise to and interacts with the electromagnetic force, one of the four fundamental forces of nature. The most familiar carriers of electrical charge are the electron and proton. Experiment has shown charge to be a conserved quantity, that is, the net charge within an isolated system will always remain constant regardless of any changes taking place within that system. Within the system, charge may be transferred between bodies, either by direct contact, or by passing along a conducting material, such as a wire. The informal term static electricity refers to the net presence (or 'imbalance') of charge on a body, usually caused when dissimilar materials are rubbed together, transferring charge from one to the other.

The charge on electrons and protons is opposite in sign, hence an amount of charge may be expressed as being either negative or positive. By convention, the charge carried by electrons is deemed negative, and that by protons positive, a custom that originated with the work of Benjamin Franklin. The amount of charge is usually given the symbol Q and expressed in coulombs; each electron carries the same charge of approximately -1.6022×10^{-19} coulomb. The proton has a charge that is equal and opposite, and thus $+1.6022\times10^{-19}$ coulomb. Charge is possessed not just by matter, but also by antimatter, each antiparticle bearing an equal and opposite charge to its corresponding particle.

Charge can be measured by a number of means, an early instrument being the gold-leaf electroscope, which although still in use for classroom demonstrations, has been superseded by the electronic electrometer.

Electric Current

The movement of electric charge is known as an electric current, the intensity of which is usually measured in amperes. Current can consist of any moving charged particles; most commonly these are electrons, but any charge in motion constitutes a current.

By historical convention, a positive current is defined as having the same direction of flow as any positive charge it contains, or to flow from the most positive part of a circuit to the most negative part. Current defined in this manner is called conventional current. The motion of negatively charged electrons around an electric circuit, one of the most familiar forms of current, is thus deemed positive in the *opposite* direction to that of the electrons. However, depending on the conditions, an electric current can consist of a flow of charged particles in either direction, or even in both directions at once. The positive-to-negative convention is widely used to simplify this situation.

An electric arc provides an energetic demonstration of electric current

The process by which electric current passes through a material is termed electrical conduction, and its nature varies with that of the charged particles and the material through which they are travelling. Examples of electric currents include metallic conduction, where electrons flow through a conductor such as metal, and electrolysis, where ions (charged atoms) flow through liquids, or through plasmas such as electrical sparks. While the particles themselves can move quite slowly, sometimes with an average drift velocity only fractions of a millimetre per second, the electric field that drives them itself propagates at close to the speed of light, enabling electrical signals to pass rapidly along wires.

Current causes several observable effects, which historically were the means of recognising its presence. That water could be decomposed by the current from a voltaic pile was discovered by Nicholson and Carlisle in 1800, a process now known as electrolysis. Their work was greatly expanded upon by Michael Faraday in 1833. Current through a resistance causes localised heating, an effect James Prescott Joule studied mathematically in 1840. One of the most important discoveries relating to current was made accidentally by Hans Christian Ørsted in 1820, when, while preparing a lecture, he witnessed the current in a wire disturbing the needle of a magnetic compass. He had discovered electromagnetism, a fundamental interaction between electricity and magnetics. The level of electromagnetic emissions generated by electric arcing is high enough to produce electromagnetic interference, which can be detrimental to the workings of adjacent equipment.

In engineering or household applications, current is often described as being either direct current (DC) or alternating current (AC). These terms refer to how the current varies in time. Direct current, as produced by example from a battery and required by most electronic devices, is a unidirectional flow from the positive part of a circuit to the negative. If, as is most common, this flow is carried by electrons, they will be travelling in the opposite direction. Alternating current is any current that reverses direction repeatedly; almost always this takes the form of a sine wave. Alternating current thus pulses back and forth within a conductor without the charge moving any net distance over time. The time-averaged value of an alternating current is zero, but it delivers energy in first one direction, and then the reverse. Alternating current is affected by electrical properties that are not observed under steady state direct current, such as inductance and capacitance. These properties however can become important when circuitry is subjected to transients, such as when first energised.

Electric Field

The concept of the electric field was introduced by Michael Faraday. An electric field is created by a charged body in the space that surrounds it, and results in a force exerted on any other charges placed within the field. The electric field acts between two charges in a similar manner to the way that the gravitational field acts between two masses, and like it, extends towards infinity and shows an inverse square relationship with distance. However, there is an important difference. Gravity always acts in attraction, drawing two masses together, while the electric field can result in either attraction or repulsion. Since large bodies such as planets generally carry no net charge, the electric field at a distance is usually zero. Thus gravity is the dominant force at distance in the universe, despite being much weaker.

An electric field generally varies in space, and its strength at any one point is defined as the force (per unit charge) that would be felt by a stationary, negligible charge if placed at that point. The conceptual charge, termed a 'test charge', must be vanishingly small to prevent its own electric field disturbing the main field and must also be stationary to prevent the effect of magnetic fields.

As the electric field is defined in terms of force, and force is a vector, so it follows that an electric field is also a vector, having both magnitude and direction. Specifically, it is a vector field.

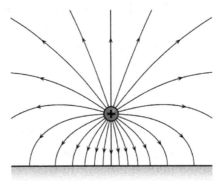

Field lines emanating from a positive charge above a plane conductor

The study of electric fields created by stationary charges is called electrostatics. The field may be visualised by a set of imaginary lines whose direction at any point is the same as that of the field. This concept was introduced by Faraday, whose term 'lines of force' still sometimes sees use. The field lines are the paths that a point positive charge would seek to make as it was forced to move within the field; they are however an imaginary concept with no physical existence, and the field permeates all the intervening space between the lines. Field lines emanating from stationary charges have several key properties: first, that they originate at positive charges and terminate at negative charges; second, that they must enter any good conductor at right angles, and third, that they may never cross nor close in on themselves.

A hollow conducting body carries all its charge on its outer surface. The field is therefore zero at all places inside the body. This is the operating principal of the Faraday cage, a conducting metal shell which isolates its interior from outside electrical effects.

The principles of electrostatics are important when designing items of high-voltage equipment. There is a finite limit to the electric field strength that may be withstood by any medium. Beyond this point, electrical breakdown occurs and an electric arc causes flashover between the charged parts. Air, for example, tends to arc across small gaps at electric field strengths which exceed 30 kV per centimetre. Over larger gaps, its breakdown strength is weaker, perhaps 1 kV per centimetre. The most visible natural occurrence of this is lightning, caused when charge becomes separated in the clouds by rising columns of air, and raises the electric field in the air to greater than it can withstand. The voltage of a large lightning cloud may be as high as 100 MV and have discharge energies as great as 250 kWh.

The field strength is greatly affected by nearby conducting objects, and it is particularly intense when it is forced to curve around sharply pointed objects. This principle is exploited in the lightning conductor, the sharp spike of which acts to encourage the lightning stroke to develop there, rather than to the building it serves to protect

Electric Potential

The concept of electric potential is closely linked to that of the electric field. A small charge placed within an electric field experiences a force, and to have brought that charge to that point against the force requires work. The electric potential at any point is defined as the energy required to bring a unit test charge from an infinite distance slowly to that point. It is usually measured in volts, and one volt is the potential for which one joule of work must be expended to bring a charge of one coulomb from infinity. This definition of potential, while formal, has little practical application, and a more useful concept is that of electric potential difference, and is the energy required to move a unit charge between two specified points. An electric field has the special property that it is *conservative*, which means that the path taken by the test charge is irrelevant: all paths between two specified points expend the same energy, and thus a unique value for potential difference may be stated. The volt is so strongly identified as the unit of choice for measurement and description of electric potential difference that the term voltage sees greater everyday usage.

A pair of AA cells. The + sign indicates the polarity of the potential difference between the battery terminals.

For practical purposes, it is useful to define a common reference point to which potentials may be expressed and compared. While this could be at infinity, a much more useful reference is the Earth itself, which is assumed to be at the same potential everywhere. This reference point naturally takes the name earth or ground. Earth is assumed to be an infinite source of equal amounts of positive and negative charge, and is therefore electrically uncharged—and unchargeable.

Electric potential is a scalar quantity, that is, it has only magnitude and not direction. It may be viewed as analogous to height: just as a released object will fall through a difference in heights caused by a gravitational field, so a charge will 'fall' across the voltage caused by an electric field. As relief maps show contour lines marking points of equal height, a set of lines marking points of equal potential (known as equipotentials) may be drawn around an electrostatically charged object. The equipotentials cross all lines of force at right angles. They must also lie parallel to a conductor's surface, otherwise this would produce a force that will move the charge carriers to even the potential of the surface.

The electric field was formally defined as the force exerted per unit charge, but the concept of potential allows for a more useful and equivalent definition: the electric field is the local gradient of the electric potential. Usually expressed in volts per metre, the vector direction of the field is the line of greatest slope of potential, and where the equipotentials lie closest together.

Electromagnets

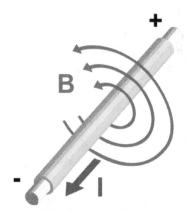

Magnetic field circles around a current

Ørsted›s discovery in 1821 that a magnetic field existed around all sides of a wire carrying an electric current indicated that there was a direct relationship between electricity and magnetism. Moreover, the interaction seemed different from gravitational and electrostatic forces, the two forces of nature then known. The force on the compass needle did not direct it to or away from the current-carrying wire, but acted at right angles to it. Ørsted's slightly obscure words were that "the electric conflict acts in a revolving manner." The force also depended on the direction of the current, for if the flow was reversed, then the force did too.

Ørsted did not fully understand his discovery, but he observed the effect was reciprocal: a current exerts a force on a magnet, and a magnetic field exerts a force on a current. The phenomenon was further investigated by Ampère, who discovered that two parallel current-carrying wires exerted a force upon each other: two wires conducting currents in the same direction are attracted to each other, while wires containing currents in opposite directions are forced apart. The interaction is mediated by the magnetic field each current produces and forms the basis for the international definition of the ampere.

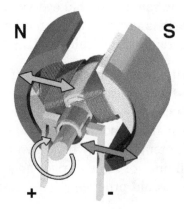

The electric motor exploits an important effect of electromagnetism: a current through a magnetic field experiences a force at right angles to both the field and current

This relationship between magnetic fields and currents is extremely important, for it led to Michael Faraday's invention of the electric motor in 1821. Faraday's homopolar motor consisted of a permanent magnet sitting in a pool of mercury. A current was allowed through a wire suspended from a pivot above the magnet and dipped into the mercury. The magnet exerted a tangential force on the wire, making it circle around the magnet for as long as the current was maintained.

Experimentation by Faraday in 1831 revealed that a wire moving perpendicular to a magnetic field developed a potential difference between its ends. Further analysis of this process, known as electromagnetic induction, enabled him to state the principle, now known as Faraday's law of induction, that the potential difference induced in a closed circuit is proportional to the rate of change of magnetic flux through the loop. Exploitation of this discovery enabled him to invent the first electrical generator in 1831, in which he converted the mechanical energy of a rotating copper disc to electrical energy. Faraday's disc was inefficient and of no use as a practical generator, but it showed the possibility of generating electric power using magnetism, a possibility that would be taken up by those that followed on from his work.

Electrochemistry

Italian physicist Alessandro Volta showing his *"battery"* to French emperor Napoleon Bonaparte in the early 19th century.

The ability of chemical reactions to produce electricity, and conversely the ability of electricity to drive chemical reactions has a wide array of uses.

Electrochemistry has always been an important part of electricity. From the initial invention of the Voltaic pile, electrochemical cells have evolved into the many different types of batteries, electroplating and electrolysis cells. Aluminium is produced in vast quantities this way, and many portable devices are electrically powered using rechargeable cells.

Electric Circuits

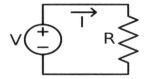

A basic electric circuit. The voltage source V on the left drives a current I around the circuit, delivering electrical energy into the resistor R. From the resistor, the current returns to the source, completing the circuit.

An electric circuit is an interconnection of electric components such that electric charge is made to flow along a closed path (a circuit), usually to perform some useful task.

The components in an electric circuit can take many forms, which can include elements such as resistors, capacitors, switches, transformers and electronics. Electronic circuits contain active components, usually semiconductors, and typically exhibit non-linear behaviour, requiring complex analysis. The simplest electric components are those that are termed passive and linear: while they may temporarily store energy, they contain no sources of it, and exhibit linear responses to stimuli.

The resistor is perhaps the simplest of passive circuit elements: as its name suggests, it resists the current through it, dissipating its energy as heat. The resistance is a consequence of the motion of charge through a conductor: in metals, for example, resistance is primarily due to collisions between electrons and ions. Ohm's law is a basic law of circuit theory, stating that the current passing through a resistance is directly proportional to the potential difference across it. The resistance of most materials is relatively constant over a range of temperatures and currents; materials under these conditions are known as 'ohmic'. The ohm, the unit of resistance, was named in honour of Georg Ohm, and is symbolised by the Greek letter Ω. 1 Ω is the resistance that will produce a potential difference of one volt in response to a current of one amp.

The capacitor is a development of the Leyden jar and is a device that can store charge, and thereby storing electrical energy in the resulting field. It consists of two conducting plates separated by a thin insulating dielectric layer; in practice, thin metal foils are coiled together, increasing the surface area per unit volume and therefore the capacitance. The unit of capacitance is the farad, named after Michael Faraday, and given the symbol F: one farad is the capacitance that develops a potential difference of one volt when it stores a charge of one coulomb. A capacitor connected to a voltage supply initially causes a current as it accumulates charge; this current will however decay in time as the capacitor fills, eventually falling to zero. A capacitor will therefore not permit a steady state current, but instead blocks it.

The inductor is a conductor, usually a coil of wire, that stores energy in a magnetic field in response to the current through it. When the current changes, the magnetic field does too, inducing a voltage between the ends of the conductor. The induced voltage is proportional to the time rate of change of the current. The constant of proportionality is termed the inductance. The unit of inductance is the henry, named after Joseph Henry, a contemporary of Faraday. One henry is the inductance that will induce a potential difference of one volt if the current through it changes at a rate of one ampere per second. The inductor's behaviour is in some regards converse to that of the capacitor: it will freely allow an unchanging current, but opposes a rapidly changing one.

Electric Power

Electric power is the rate at which electric energy is transferred by an electric circuit. The SI unit of power is the watt, one joule per second.

Electric power, like mechanical power, is the rate of doing work, measured in watts, and represented by the letter P. The term *wattage* is used colloquially to mean "electric power in watts." The electric power in watts produced by an electric current I consisting of a charge of Q coulombs every t seconds passing through an electric potential (voltage) difference of V is

$$P = \text{work done per unit time} = \frac{QV}{t} = IV$$

where

Q is electric charge in coulombs

t is time in seconds

I is electric current in amperes

V is electric potential or voltage in volts

Electricity generation is often done with electric generators, but can also be supplied by chemical sources such as electric batteries or by other means from a wide variety of sources of energy. Electric power is generally supplied to businesses and homes by the electric power industry. Electricity is usually sold by the kilowatt hour (3.6 MJ) which is the product of power in kilowatts multiplied by running time in hours. Electric utilities measure power using electricity meters, which keep a running total of the electric energy delivered to a customer. Unlike fossil fuels, electricity is a low entropy form of energy and can be converted into motion or many other forms of energy with high efficiency.

Electronics

Surface mount electronic components

Electronics deals with electrical circuits that involve active electrical components such as vacuum tubes, transistors, diodes and integrated circuits, and associated passive interconnection technologies. The nonlinear behaviour of active components and their ability to control electron flows makes amplification of weak signals possible and electronics is widely used in information processing, telecommunications, and signal processing. The ability of electronic devices to act as switches makes digital information processing possible. Interconnection technologies such as circuit boards, electronics packaging technology, and other varied forms of communication infrastructure complete circuit functionality and transform the mixed components into a regular working system.

Today, most electronic devices use semiconductor components to perform electron control. The study of semiconductor devices and related technology is considered a branch of solid state physics, whereas the design and construction of electronic circuits to solve practical problems come under electronics engineering.

Electromagnetic Wave

Faraday's and Ampère's work showed that a time-varying magnetic field acted as a source of an electric field, and a time-varying electric field was a source of a magnetic field. Thus, when either field is changing in time, then a field of the other is necessarily induced. Such a phenomenon has the properties of a wave, and is naturally referred to as an electromagnetic wave. Electromagnetic waves were analysed theoretically by James Clerk Maxwell in 1864. Maxwell developed a set of equations that could unambiguously describe the interrelationship between electric field, magnetic field, electric charge, and electric current. He could moreover prove that such a wave would necessarily travel at the speed of light, and thus light itself was a form of electromagnetic radiation. Maxwell's Laws, which unify light, fields, and charge are one of the great milestones of theoretical physics.

Thus, the work of many researchers enabled the use of electronics to convert signals into high frequency oscillating currents, and via suitably shaped conductors, electricity permits the transmission and reception of these signals via radio waves over very long distances.

Production and Uses

Generation and Transmission

Early 20th-century alternator made in Budapest, Hungary, in the power generating hall of a hydroelectric station (photograph by Prokudin-Gorsky, 1905–1915).

In the 6th century BC, the Greek philosopher Thales of Miletus experimented with amber rods and these experiments were the first studies into the production of electrical energy. While this method, now known as the triboelectric effect, can lift light objects and generate sparks, it is extremely inefficient. It was not until the invention of the voltaic pile in the eighteenth century that a viable source of electricity became available. The voltaic pile, and its modern descendant, the electrical battery, store energy chemically and make it available on demand in the form of electrical energy. The battery is a versatile and very common power source which is ideally suited to many applications, but its energy storage is finite, and once discharged it must be disposed of or recharged. For large electrical demands electrical energy must be generated and transmitted continuously over conductive transmission lines.

Electrical power is usually generated by electro-mechanical generators driven by steam produced from fossil fuel combustion, or the heat released from nuclear reactions; or from other sources such as kinetic energy extracted from wind or flowing water. The modern steam turbine invented by Sir Charles Parsons in 1884 today generates about 80 percent of the electric power in the world using a variety of heat sources. Such generators bear no resemblance to Faraday's homopolar disc generator of 1831, but they still rely on his electromagnetic principle that a conductor linking a changing magnetic field induces a potential difference across its ends. The invention in the late nineteenth century of the transformer meant that electrical power could be transmitted more efficiently at a higher voltage but lower current. Efficient electrical transmission meant in turn that electricity could be generated at centralised power stations, where it benefited from economies of scale, and then be despatched relatively long distances to where it was needed.

Wind power is of increasing importance in many countries

Since electrical energy cannot easily be stored in quantities large enough to meet demands on a national scale, at all times exactly as much must be produced as is required. This requires electricity utilities to make careful predictions of their electrical loads, and maintain constant co-ordination with their power stations. A certain amount of generation must always be held in reserve to cushion an electrical grid against inevitable disturbances and losses.

Demand for electricity grows with great rapidity as a nation modernises and its economy develops. The United States showed a 12% increase in demand during each year of the first three decades of the twentieth century, a rate of growth that is now being experienced by emerging economies such as those of India or China. Historically, the growth rate for electricity demand has outstripped that for other forms of energy.

Environmental concerns with electricity generation have led to an increased focus on generation from renewable sources, in particular from wind and hydropower. While debate can be expected to continue over the environmental impact of different means of electricity production, its final form is relatively clean

Applications

Electricity is a very convenient way to transfer energy, and it has been adapted to a huge, and growing, number of uses. The invention of a practical incandescent light bulb in the 1870s led to lighting becoming one of the first publicly available applications of electrical power. Although electrification brought with it its own dangers, replacing the naked flames of gas lighting greatly reduced fire hazards within homes and factories. Public utilities were set up in many cities targeting the burgeoning market for electrical lighting.

The light bulb, an early application of electricity, operates by Joule heating: the passage of current through resistance generating heat

The resistive Joule heating effect employed in filament light bulbs also sees more direct use in electric heating. While this is versatile and controllable, it can be seen as wasteful, since most electrical generation has already required the production of heat at a power station. A number of countries, such as Denmark, have issued legislation restricting or banning the use of resistive electric heating in new buildings. Electricity is however still a highly practical energy source for heating and refrigeration, with air conditioning/heat pumps representing a growing sector for electricity demand for heating and cooling, the effects of which electricity utilities are increasingly obliged to accommodate.

Electricity is used within telecommunications, and indeed the electrical telegraph, demonstrated commercially in 1837 by Cooke and Wheatstone, was one of its earliest applications. With the construction of first intercontinental, and then transatlantic, telegraph systems in the 1860s, electricity had enabled communications in minutes across the globe. Optical fibre and satellite communication have taken a share of the market for communications systems, but electricity can be expected to remain an essential part of the process.

The effects of electromagnetism are most visibly employed in the electric motor, which provides a clean and efficient means of motive power. A stationary motor such as a winch is easily provided with a supply of power, but a motor that moves with its application, such as an electric vehicle, is obliged to either carry along a power source such as a battery, or to collect current from a sliding contact such as a pantograph.

Electronic devices make use of the transistor, perhaps one of the most important inventions of the twentieth century, and a fundamental building block of all modern circuitry. A modern integrated circuit may contain several billion miniaturised transistors in a region only a few centimetres square.

Electricity is also used to fuel public transportation, including electric buses and trains.

Electricity and the Natural World

Physiological Effects

A voltage applied to a human body causes an electric current through the tissues, and although the relationship is non-linear, the greater the voltage, the greater the current. The threshold for perception varies with the supply frequency and with the path of the current, but is about 0.1 mA to 1 mA for mains-frequency electricity, though a current as low as a microamp can be detected as an electrovibration effect under certain conditions. If the current is sufficiently high, it will cause muscle contraction, fibrillation of the heart, and tissue burns. The lack of any visible sign that a conductor is electrified makes electricity a particular hazard. The pain caused by an electric shock can be intense, leading electricity at times to be employed as a method of torture. Death caused by an electric shock is referred to as electrocution. Electrocution is still the means of judicial execution in some jurisdictions, though its use has become rarer in recent times.

Electrical Phenomena in Nature

The electric eel, *Electrophorus electricus*

Electricity is not a human invention, and may be observed in several forms in nature, a prominent manifestation of which is lightning. Many interactions familiar at the macroscopic level, such as touch, friction or chemical bonding, are due to interactions between electric fields on the atomic scale. The Earth's magnetic field is thought to arise from a natural dynamo of circulating currents in the planet's core. Certain crystals, such as quartz, or even sugar, generate a potential difference across their faces when subjected to external pressure. The effect is reciprocal, and when a piezoelectric material is subjected to an electric field, a small change in physical dimensions takes place.

Some organisms, such as sharks, are able to detect and respond to changes in electric fields, an ability known as electroreception, while others, termed electrogenic, are able to generate voltages themselves to serve as a predatory or defensive weapon. The order Gymnotiformes, of which the best known example is the electric eel, detect or stun their prey via high voltages generated from modified muscle cells called electrocytes. All animals transmit information along their cell membranes with voltage pulses called action potentials, whose functions include communication by the nervous system between neurons and muscles. An electric shock stimulates this system, and causes muscles to contract. Action potentials are also responsible for coordinating activities in certain plants.

Cultural Perception

In 1850, William Gladstone asked the scientist Michael Faraday why electricity was valuable. Faraday answered, "One day sir, you may tax it."

In the 19th and early 20th century, electricity was not part of the everyday life of many people, even in the industrialised Western world. The popular culture of the time accordingly often depicts it as a mysterious, quasi-magical force that can slay the living, revive the dead or otherwise bend the laws of nature. This attitude began with the 1771 experiments of Luigi Galvani in which the legs of dead frogs were shown to twitch on application of animal electricity. "Revitalization" or resuscitation of apparently dead or drowned persons was reported in the medical literature shortly after Galvani's work. These results were known to Mary Shelley when she authored *Frankenstein* (1819), although she does not name the method of revitalization of the monster. The revitalization of monsters with electricity later became a stock theme in horror films.

As the public familiarity with electricity as the lifeblood of the Second Industrial Revolution grew, its wielders were more often cast in a positive light, such as the workers who "finger death at their gloves' end as they piece and repiece the living wires" in Rudyard Kipling's 1907 poem *Sons of Martha*. Electrically powered vehicles of every sort featured large in adventure stories such as those of Jules Verne and the *Tom Swift* books. The masters of electricity, whether fictional or real—including scientists such as Thomas Edison, Charles Steinmetz or Nikola Tesla—were popularly conceived of as having wizard-like powers.

With electricity ceasing to be a novelty and becoming a necessity of everyday life in the later half of the 20th century, it required particular attention by popular culture only when it *stops* flowing, an event that usually signals disaster. The people who *keep* it flowing, such as the nameless hero of Jimmy Webb's song "Wichita Lineman" (1968), are still often cast as heroic, wizard-like figures.

Electronics

Surface-mount electronic components

Electronics is the science of controlling electric energy, energy in which the electrons have a fundamental role. Electronics deals with electrical circuits that involve active electrical components such as vacuum tubes, transistors, diodes and integrated circuits, and associated passive electrical components and interconnection technologies. Commonly, electronic devices contain circuitry consisting primarily or exclusively of active semiconductors supplemented with passive elements; such a circuit is described as an electronic circuit.

The nonlinear behaviour of active components and their ability to control electron flows makes amplification of weak signals possible, and electronics is widely used in information processing, telecommunication, and signal processing. The ability of electronic devices to act as switches makes digital information processing possible. Interconnection technologies such as circuit boards, electronics packaging technology, and other varied forms of communication infrastructure complete circuit functionality and transform the mixed components into a regular working system.

Electronics is distinct from electrical and electro-mechanical science and technology, which deal with the generation, distribution, switching, storage, and conversion of electrical energy to and from other energy forms using wires, motors, generators, batteries, switches, relays, transformers, resistors, and other passive components. This distinction started around 1906 with the invention by Lee De Forest of the triode, which made electrical amplification of weak radio signals and audio signals possible with a non-mechanical device. Until 1950 this field was called "radio technology" because its principal application was the design and theory of radio transmitters, receivers, and vacuum tubes.

Today, most electronic devices use semiconductor components to perform electron control. The study of semiconductor devices and related technology is considered a branch of solid-state physics, whereas the design and construction of electronic circuits to solve practical problems come under electronics engineering.

Branches of Electronics

Electronics has branches as follows:

1. Digital electronics

2. Analogue electronics

3. Microelectronics

4. Circuit design

5. Integrated circuits

6. Optoelectronics

7. Semiconductor devices

8. Embedded systems

Electronic Devices and Components

Electronics Technician performing a voltage check on a power circuit card in the air navigation equipment room aboard the aircraft carrier USS *Abraham Lincoln* (CVN 72).

An electronic component is any physical entity in an electronic system used to affect the electrons or their associated fields in a manner consistent with the intended function of the electronic system. Components are generally intended to be connected together, usually by being soldered to a printed circuit board (PCB), to create an electronic circuit with a particular function (for example an amplifier, radio receiver, or oscillator). Components may be packaged singly, or in more complex groups as integrated circuits. Some common electronic components are capacitors, inductors, resistors, diodes, transistors, etc. Components are often categorized as active (e.g. transistors and thyristors) or passive (e.g. resistors, diodes, inductors and capacitors).

History of Electronic Components

Vacuum tubes (Thermionic valves) were one of the earliest electronic components. They were almost solely responsible for the electronics revolution of the first half of the Twentieth Century. They took electronics from parlor tricks and gave us radio, television, phonographs, radar, long distance telephony and much more. They played a leading role in the field of microwave and high power transmission as well as television receivers until the middle of the 1980s. Since that time, solid state devices have all but completely taken over. Vacuum tubes are still used in some specialist applications such as high power RF amplifiers, cathode ray tubes, specialist audio equipment, guitar amplifiers and some microwave devices.

In April 1955 the IBM 608 was the first IBM product to use transistor circuits without any vacuum tubes and is believed to be the world's first all-transistorized calculator to be manufactured for the commercial market. The 608 contained more than 3,000 germanium transistors. Thomas J. Watson Jr. ordered all future IBM products to use transistors in their design. From that time on transistors were almost exclusively used for computer logic and peripherals.

Types of Circuits

Circuits and components can be divided into two groups: analog and digital. A particular device may consist of circuitry that has one or the other or a mix of the two types.

Analog Circuits

Hitachi J100 adjustable frequency drive chassis

Most analog electronic appliances, such as radio receivers, are constructed from combinations of a few types of basic circuits. Analog circuits use a continuous range of voltage or current as opposed to discrete levels as in digital circuits.

The number of different analog circuits so far devised is huge, especially because a 'circuit' can be defined as anything from a single component, to systems containing thousands of components.

Analog circuits are sometimes called linear circuits although many non-linear effects are used in analog circuits such as mixers, modulators, etc. Good examples of analog circuits include vacuum tube and transistor amplifiers, operational amplifiers and oscillators.

One rarely finds modern circuits that are entirely analog. These days analog circuitry may use digital or even microprocessor techniques to improve performance. This type of circuit is usually called "mixed signal" rather than analog or digital.

Sometimes it may be difficult to differentiate between analog and digital circuits as they have elements of both linear and non-linear operation. An example is the comparator which takes in a continuous range of voltage but only outputs one of two levels as in a digital circuit. Similarly, an overdriven transistor amplifier can take on the characteristics of a controlled switch having essentially two levels of output. In fact, many digital circuits are actually implemented as variations of analog circuits similar to this example—after all, all aspects of the real physical world are essentially analog, so digital effects are only realized by constraining analog behavior.

Digital Circuits

Digital circuits are electric circuits based on a number of discrete voltage levels. Digital circuits are the most common physical representation of Boolean algebra, and are the basis of all digital computers. To most engineers, the terms "digital circuit", "digital system" and "logic" are interchangeable in the context of digital circuits. Most digital circuits use a binary system with two voltage levels labeled "0" and "1". Often logic "0" will be a lower voltage and referred to as "Low" while logic "1" is referred to as "High". However, some systems use the reverse definition ("0" is "High") or are current based. Quite often the logic designer may reverse these definitions from one circuit to the next as he sees fit to facilitate his design. The definition of the levels as "0" or "1" is arbitrary.

Ternary (with three states) logic has been studied, and some prototype computers made.

Computers, electronic clocks, and programmable logic controllers (used to control industrial processes) are constructed of digital circuits. Digital signal processors are another example.

Building blocks:

- Logic gates
- Adders
- Flip-flops
- Counters

- Registers
- Multiplexers
- Schmitt triggers

Highly integrated devices:

- Microprocessors
- Microcontrollers
- Application-specific integrated circuit (ASIC)
- Digital signal processor (DSP)
- Field-programmable gate array (FPGA)

Heat Dissipation and Thermal Management

Heat generated by electronic circuitry must be dissipated to prevent immediate failure and improve long term reliability. Heat dissipation is mostly achieved by passive conduction/convection. Means to achieve greater dissipation include heat sinks and fans for air cooling, and other forms of computer cooling such as water cooling. These techniques use convection, conduction, and radiation of heat energy.

Noise

Electronic noise is defined as unwanted disturbances superposed on a useful signal that tend to obscure its information content. Noise is not the same as signal distortion caused by a circuit. Noise is associated with all electronic circuits. Noise may be electromagnetically or thermally generated, which can be decreased by lowering the operating temperature of the circuit. Other types of noise, such as shot noise cannot be removed as they are due to limitations in physical properties.

Electronics Theory

Mathematical methods are integral to the study of electronics. To become proficient in electronics it is also necessary to become proficient in the mathematics of circuit analysis.

Circuit analysis is the study of methods of solving generally linear systems for unknown variables such as the voltage at a certain node or the current through a certain branch of a network. A common analytical tool for this is the SPICE circuit simulator.

Also important to electronics is the study and understanding of electromagnetic field theory.

Electronics Lab

Due to the complex nature of electronics theory, laboratory experimentation is an important part of the development of electronic devices. These experiments are used to test or verify the engineer's design and detect errors. Historically, electronics labs have consisted of electronics devices and equipment located in a physical space, although in more recent years the trend has been towards electronics lab simulation software, such as CircuitLogix, Multisim, and PSpice.

Computer Aided Design (CAD)

Today's electronics engineers have the ability to design circuits using premanufactured building blocks such as power supplies, semiconductors (i.e. semiconductor devices, such as transistors), and integrated circuits. Electronic design automation software programs include schematic capture programs and printed circuit board design programs. Popular names in the EDA software world are NI Multisim, Cadence (ORCAD), EAGLE PCB and Schematic, Mentor (PADS PCB and LOGIC Schematic), Altium (Protel), LabCentre Electronics (Proteus), gEDA, KiCad and many others.

Construction Methods

Many different methods of connecting components have been used over the years. For instance, early electronics often used point to point wiring with components attached to wooden breadboards to construct circuits. Cordwood construction and wire wrap were other methods used. Most modern day electronics now use printed circuit boards made of materials such as FR4, or the cheaper (and less hard-wearing) Synthetic Resin Bonded Paper (SRBP, also known as Paxoline/Paxolin (trade marks) and FR2) - characterised by its brown colour. Health and environmental concerns associated with electronics assembly have gained increased attention in recent years, especially for products destined to the European Union, with its Restriction of Hazardous Substances Directive (RoHS) and Waste Electrical and Electronic Equipment Directive (WEEE), which went into force in July 2006.

Degradation

Rasberry crazy ants have been known to consume the insides of electrical wiring, and nest inside of electronics; they prefer DC to AC currents. This behavior is not well understood by scientists.

Microelectronics

Microelectronics is a subfield of electronics. As the name suggests, microelectronics relates to the study and manufacture (or microfabrication) of very small electronic designs and components. Usually, but not always, this means micrometre-scale or smaller. These devices are typically made from semiconductor materials. Many components of normal electronic design are available in a microelectronic equivalent. These include transistors, capacitors, inductors, resistors, diodes and (naturally) insulators and conductors can all be found in microelectronic devices. Unique wiring techniques such as wire bonding are also often used in microelectronics because of the unusually small size of the components, leads and pads. This technique requires specialized equipment and is expensive.

Digital integrated circuits (ICs) consist mostly of transistors. Analog circuits commonly contain resistors and capacitors as well. Inductors are used in some high frequency analog circuits, but tend to occupy large chip area if used at low frequencies; gyrators can replace them in many applications.

As techniques improve, the scale of microelectronic components continues to decrease. At smaller scales, the relative impact of intrinsic circuit properties such as interconnections may become more significant. These are called parasitic effects, and the goal of the microelectronics design engineer is to find ways to compensate for or to minimize these effects, while always delivering smaller, faster, and cheaper devices.

Electromagnetism

Electromagnetism is a branch of physics which involves the study of the electromagnetic force, a type of physical interaction that occurs between electrically charged particles. The electromagnetic force usually exhibits electromagnetic fields, such as electric fields, magnetic fields, and light. The electromagnetic force is one of the four fundamental interactions (commonly called forces) in nature. The other three fundamental interactions are the strong interaction, the weak interaction, and gravitation.

Lightning is an electrostatic discharge that travels between two charged regions.

Electromagnetic phenomena are defined in terms of the electromagnetic force, sometimes called the Lorentz force, which includes both electricity and magnetism as different manifestations of the same phenomenon.

The electromagnetic force plays a major role in determining the internal properties of most objects encountered in daily life. Ordinary matter takes its form as a result of intermolecular forces between individual atoms and molecules in matter, and are a manifestation of the electromagnetic force. Electrons are bound by the electromagnetic force to atomic nuclei, and their orbital shapes and their influence on nearby atoms with their electrons is described by quantum mechanics. The electromagnetic force governs the processes involved in chemistry, which arise from interactions between the electrons of neighboring atoms.

There are numerous mathematical descriptions of the electromagnetic field. In classical electrodynamics, electric fields are described as electric potential and electric current. In Faraday's law, magnetic fields are associated with electromagnetic induction and magnetism, and Maxwell's equations describe how electric and magnetic fields are generated and altered by each other and by charges and currents.

The theoretical implications of electromagnetism, in particular the establishment of the speed of light based on properties of the "medium" of propagation (permeability and permittivity), led to the development of special relativity by Albert Einstein in 1905.

Although electromagnetism is considered one of the four fundamental forces, at high energy the weak force and electromagnetic force are unified as a single electroweak force. In the history of the universe, during the quark epoch the unified force broke into the two separate forces as the universe cooled.

History of the Theory

Hans Christian Ørsted.

Originally, electricity and magnetism were thought of as two separate forces. This view changed, however, with the publication of James Clerk Maxwell's 1873 *A Treatise on Electricity and Magnetism* in which the interactions of positive and negative charges were shown to be mediated by one force. There are four main effects resulting from these interactions, all of which have been clearly demonstrated by experiments:

1. Electric charges attract or repel one another with a force inversely proportional to the square of the distance between them: unlike charges attract, like ones repel.

2. Magnetic poles (or states of polarization at individual points) attract or repel one another in a manner similar to positive and negative charges and always exist as pairs: every north pole is yoked to a south pole.

3. An electric current inside a wire creates a corresponding circumferential magnetic field outside the wire. Its direction (clockwise or counter-clockwise) depends on the direction of the current in the wire.

4. A current is induced in a loop of wire when it is moved toward or away from a magnetic field, or a magnet is moved towards or away from it; the direction of current depends on that of the movement.

André-Marie Ampère

While preparing for an evening lecture on 21 April 1820, Hans Christian Ørsted made a surprising observation. As he was setting up his materials, he noticed a compass needle deflected away from magnetic north when the electric current from the battery he was using was switched on and off. This deflection convinced him that magnetic fields radiate from all sides of a wire carrying an electric current, just as light and heat do, and that it confirmed a direct relationship between electricity and magnetism.

Michael Faraday

At the time of discovery, Ørsted did not suggest any satisfactory explanation of the phenomenon, nor did he try to represent the phenomenon in a mathematical framework. However, three months later he began more intensive investigations. Soon thereafter he published his findings, proving that an electric current produces a magnetic field as it flows through a wire. The CGS unit of magnetic induction (oersted) is named in honor of his contributions to the field of electromagnetism.

James Clerk Maxwell

His findings resulted in intensive research throughout the scientific community in electrodynamics. They influenced French physicist André-Marie Ampère's developments of a single mathematical form to represent the magnetic forces between current-carrying conductors. Ørsted's discovery also represented a major step toward a unified concept of energy.

This unification, which was observed by Michael Faraday, extended by James Clerk Maxwell, and partially reformulated by Oliver Heaviside and Heinrich Hertz, is one of the key accomplishments of 19th century mathematical physics. It had far-reaching consequences, one of which was the understanding of the nature of light. Unlike what was proposed by electromagnetic theory of that time, light and other electromagnetic waves are at present seen as taking the form of quantized, self-propagating oscillatory electromagnetic field disturbances called photons. Different frequencies of oscillation give rise to the different forms of electromagnetic radiation, from radio waves at the lowest frequencies, to visible light at intermediate frequencies, to gamma rays at the highest frequencies.

Ørsted was not the only person to examine the relationship between electricity and magnetism. In 1802, Gian Domenico Romagnosi, an Italian legal scholar, deflected a magnetic needle using electrostatic charges. Actually, no galvanic current existed in the setup and hence no electromagnetism was present. An account of the discovery was published in 1802 in an Italian newspaper, but it was largely overlooked by the contemporary scientific community.

Fundamental Forces

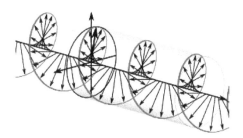

Representation of the electric field vector of a wave of circularly polarized electromagnetic radiation.

The electromagnetic force is one of the four known fundamental forces. The other fundamental forces are:

- the weak nuclear force, which binds to all known particles in the Standard Model, and causes certain forms of radioactive decay. (In particle physics though, the electroweak interaction is the unified description of two of the four known fundamental interactions of nature: electromagnetism and the weak interaction);

- the strong nuclear force, which binds quarks to form nucleons, and binds nucleons to form nuclei

- the gravitational force.

All other forces (e.g., friction, contact forces) are derived from these four fundamental forces (including momentum which is carried by the movement of particles).

The electromagnetic force is the one responsible for practically all the phenomena one encounters in daily life above the nuclear scale, with the exception of gravity. Roughly speaking, all the forces involved in interactions between atoms can be explained by the electromagnetic force acting between the electrically charged atomic nuclei and electrons of the atoms. Electromagnetic forces also explain how these particles carry momentum by their movement. This includes the forces we experience in "pushing" or "pulling" ordinary material objects, which result from the intermolecular forces that act between the individual molecules in our bodies and those in the objects. The electromagnetic force is also involved in all forms of chemical phenomena.

A necessary part of understanding the intra-atomic and intermolecular forces is the effective force generated by the momentum of the electrons' movement, such that as electrons move between interacting atoms they carry momentum with them. As a collection of electrons becomes more confined, their minimum momentum necessarily increases due to the Pauli exclusion principle. The behaviour of matter at the molecular scale including its density is determined by the balance between the electromagnetic force and the force generated by the exchange of momentum carried by the electrons themselves.

Classical Electrodynamics

The scientist William Gilbert proposed, in his *De Magnete* (1600), that electricity and magnetism, while both capable of causing attraction and repulsion of objects, were distinct effects. Mariners had noticed that lightning strikes had the ability to disturb a compass needle, but the link between lightning and electricity was not confirmed until Benjamin Franklin's proposed experiments in 1752. One of the first to discover and publish a link between man-made electric current and magnetism was Romagnosi, who in 1802 noticed that connecting a wire across a voltaic pile deflected a nearby compass needle. However, the effect did not become widely known until 1820, when Ørsted performed a similar experiment. Ørsted's work influenced Ampère to produce a theory of electromagnetism that set the subject on a mathematical foundation.

A theory of electromagnetism, known as classical electromagnetism, was developed by various physicists over the course of the 19th century, culminating in the work of James Clerk Maxwell, who unified the preceding developments into a single theory and discovered the electromagnetic nature of light. In classical electromagnetism, the behavior of the electromagnetic field is described by a set of equations known as Maxwell's equations, and the electromagnetic force is given by the Lorentz force law.

One of the peculiarities of classical electromagnetism is that it is difficult to reconcile with classical mechanics, but it is compatible with special relativity. According to Maxwell's equations, the speed of light in a vacuum is a universal constant that is dependent only on the electrical permittivity and magnetic permeability of free space. This violates Galilean invariance, a long-standing cornerstone of classical mechanics. One way to reconcile the two theories (electromagnetism and classical mechanics) is to assume the existence of a luminiferous aether through which the light propagates. However, subsequent experimental efforts failed to detect the presence of the aether. After important contributions of Hendrik Lorentz and Henri Poincaré, in 1905, Albert Einstein solved the problem with the introduction of special relativity, which replaced classical kinematics with a new theory of kinematics compatible with classical electromagnetism.

In addition, relativity theory implies that in moving frames of reference a magnetic field transforms to a field with a nonzero electric component and conversely, a moving electric field transforms to a nonzero magnetic component, thus firmly showing that the phenomena are two sides of the same coin. Hence the term "electromagnetism".

Quantum Mechanics

Photoelectric Effect

In a second paper published in 1905, Albert Einstein undermined the very foundations of classical electromagnetism. In his theory of the photoelectric effect, for which he won the Nobel prize in physics, he posited that light could exist in discrete particle-like quantities, which later came to be called photons. Einstein's theory of the photoelectric effect extended the insights that appeared in the solution of the ultraviolet catastrophe presented by Max Planck in 1900 and who coined the term "quanta" . In his work, Planck showed that hot objects emit electromagnetic radiation in discrete packets ("quanta"), which leads to a finite total energy emitted as black body radiation. Both of these results were in direct contradiction with the classical view of light as a continuous wave. Planck's and Einstein's theories were progenitors of quantum mechanics, which, when formulated in 1925, necessitated the invention of a quantum theory of electromagnetism. This theory, completed in the 1940s–1950s, is known as quantum electrodynamics (or "QED"), and, in situations where perturbation theory can be applied, is one of the most accurate theories known to physics.

Quantum Electrodynamics

All electromagnetic phenomena can be explained in terms of quantum mechanics, specifically by quantum electrodynamics (which includes classical electrodynamics as a limiting case) and this accounts for almost all physical phenomena observable to the unaided human senses, including electromagnetic radiation (light), all of chemistry, most of mechanics (excepting gravitation), and, of course, magnetism and electricity.

Electroweak Interaction

The electroweak interaction is the unified field theory description of two of the four known fundamental interactions of nature: electromagnetism and the weak interaction. Although these two forces appear very different at everyday low energies, the theory models them as two different aspects of the same force. At energies greater than 100 GeV, called the unification energy, the two forces merge into a single electroweak force. Thus when the universe was hot enough (approximately 10^{15} K, a temperature that was exceeded until shortly after the Big Bang) the electromagnetic force and weak force were merged into the electroweak force. With the cooling of the universe, at the end of the electroweak epoch, the electroweak force separated into the electromagnetic force and the weak force. During the following quark epoch, it was still too hot for quarks to combine into hadrons and they moved about freely.

Quantities and Units

Electromagnetic units are part of a system of electrical units based primarily upon the magnetic properties of electric currents, the fundamental SI unit being the ampere. The units are:

- ampere (electric current)
- coulomb (electric charge)
- farad (capacitance)

- henry (inductance)
- ohm (resistance)
- siemens (conductance)
- tesla (magnetic flux density)
- volt (electric potential)
- watt (power)
- weber (magnetic flux)

In the electromagnetic cgs system, electric current is a fundamental quantity defined via Ampère's law and takes the permeability as a dimensionless quantity (relative permeability) whose value in a vacuum is unity. As a consequence, the square of the speed of light appears explicitly in some of the equations interrelating quantities in this system.

SI electromagnetism units				
Symbol	Name of Quantity	Derived Units	Unit	Base Units
I	electric current	ampere (SI base unit)	A	A (= W/V = C/s)
Q	electric charge	coulomb	C	A·s
U, ΔV, $\Delta\varphi$; E	potential difference; electromotive force	volt	V	kg·m^2·s^{-3}·A^{-1} (= J/C)
R; Z; X	electric resistance; impedance; reactance	ohm	Ω	kg·m^2·s^{-3}·A^{-2} (= V/A)
ρ	resistivity	ohm metre	Ω·m	kg·m^3·s^{-3}·A^{-2}
P	electric power	watt	W	kg·m^2·s^{-3} (= V·A)
C	capacitance	farad	F	kg^{-1}·m^{-2}·s^4·A^2 (= C/V)
E	electric field strength	volt per metre	V/m	kg·m·s^{-3}·A^{-1} (= N/C)
D	electric displacement field	coulomb per square metre	C/m^2	A·s·m^{-2}
ε	permittivity	farad per metre	F/m	kg^{-1}·m^{-3}·s^4·A^2
χ_e	electric susceptibility	(dimensionless)	–	–
G; Y; B	conductance; admittance; susceptance	siemens	S	kg^{-1}·m^{-2}·s^3·A^2 (= Ω^{-1})
κ, γ, σ	conductivity	siemens per metre	S/m	kg^{-1}·m^{-3}·s^3·A^2
B	magnetic flux density, magnetic induction	tesla	T	kg·s^{-2}·A^{-1} (= Wb/m^2 = N·A^{-1}·m^{-1})
Φ	magnetic flux	weber	Wb	kg·m^2·s^{-2}·A^{-1} (= V·s)
H	magnetic field strength	ampere per metre	A/m	A·m^{-1}
L, M	inductance	henry	H	kg·m^2·s^{-2}·A^{-2} (= Wb/A = V·s/A)
μ	permeability	henry per metre	H/m	kg·m·s^{-2}·A^{-2}
χ	magnetic susceptibility	(dimensionless)	–	–

Formulas for physical laws of electromagnetism (such as Maxwell's equations) need to be adjusted depending on what system of units one uses. This is because there is no one-to-one correspondence between electromagnetic units in SI and those in CGS, as is the case for mechanical units. Furthermore, within CGS, there are several plausible choices of electromagnetic units, leading to different unit "sub-systems", including Gaussian, "ESU", "EMU", and Heaviside–Lorentz. Among these choices, Gaussian units are the most common today, and in fact the phrase "CGS units" is often used to refer specifically to CGS-Gaussian units.

References

- International Union of Pure and Applied Chemistry (1993). Quantities, Units and Symbols in Physical Chemistry, 2nd edition, Oxford: Blackwell Science. ISBN 0-632-03583-8. pp. 14–15. Electronic version.

- Ravaioli, Fawwaz T. Ulaby, Eric Michielssen, Umberto (2010). Fundamentals of applied electromagnetics (6th ed.). Boston: Prentice Hall. p. 13. ISBN 978-0-13-213931-1.

- Pugh, Emerson W.; Johnson, Lyle R.; Palmer, John H. (1991). IBM's 360 and early 370 systems. MIT Press. p. 34. ISBN 0-262-16123-0.

- Sōgo Okamura (1994). History of Electron Tubes. IOS Press. p. 5. ISBN 978-90-5199-145-1. Retrieved 5 December 2012.

- Patterson, Walter C. (1999), Transforming Electricity: The Coming Generation of Change, Earthscan, ISBN 1-85383-341-X

- Hammond, Percy (1981), "Electromagnetism for Engineers", Nature, Pergamon, 168 (4262): 4, Bibcode:1951Natur.168....4G, doi:10.1038/168004b0, ISBN 0-08-022104-1

4

Principles and Concepts of Electrical Engineering

In order to completely understand electrical engineering, it is necessary to understand the principles and concepts of the subject. Electrical wiring is necessary for devices such as meters and light fittings whereas a daisy chain is a wiring system which has numerous gadgets connected to it in a sequence. The following section elucidates the mechanisms associated with this area of study.

Electrical Wiring

5 core PVC insulated copper cable

Building wiring is the electrical wiring and associated devices such as switches, meters and light fittings used in buildings or other structures. Electrical wiring uses insulated conductors.

Wires and cables are rated by the circuit voltage, temperature and environmental conditions (moisture, sunlight, oil, chemicals) in which they can be used, and their maximum current. Wiring safety codes vary by country, and the International Electrotechnical Commission (IEC) is attempting to standardise wiring amongst member countries. Colour codes are used to distinguish line, neutral and earth (ground) wires.

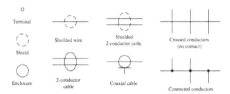

Electrical symbols for wiring

Wiring Safety Codes

Wiring safety codes are intended to protect people and property from electrical shock and fire hazards. Regulations may be established by city, county, provincial/state or national legislation, usually by adopting a model code (with or without local amendments) produced by a technical standards-setting organisation, or by a national standard electrical code.

First Electrical codes arose in the 1880s with the commercial introduction of electrical power. Many conflicting standards existed for the selection of wire sizes and other design rules for electrical installations.

The first electrical codes in the United States originated in New York in 1881 to regulate installations of electric lighting. Since 1897 the US National Fire Protection Association, a private non-profit association formed by insurance companies, has published the *National Electrical Code* (NEC). States, counties or cities often include the NEC in their local building codes by reference along with local differences. The NEC is modified every three years. It is a consensus code considering suggestions from interested parties. The proposals are studied by committees of engineers, tradesmen, manufacturer representatives, fire fighters and other invitees.

Since 1927, the Canadian Standards Association (CSA) has produced the Canadian *Safety Standard for Electrical Installations*, which is the basis for provincial electrical codes. The CSA also produces the Canadian Electrical Code, the 2006 edition of which references IEC 60364 (*Electrical Installations for Buildings*) and states that the code addresses the fundamental principles of electrical protection in Section 131. The Canadian code reprints Chapter 13 of IEC 60364, but there are no numerical criteria listed in that chapter to assess the adequacy of any electrical installation.

Although the US and Canadian national standards deal with the same physical phenomena and broadly similar objectives, they differ occasionally in technical detail. As part of the North American Free Trade Agreement (NAFTA) program, US and Canadian standards are slowly converging toward each other, in a process known as harmonisation.

In Germany, DKE (the German Commission for Electrical, Electronic and Information Technologies of DIN and VDE) is the organisation responsible for the promulgation of electrical standards and safety specifications. DIN VDE 0100 is the German wiring regulations document harmonised with IEC 60364.

In the United Kingdom, wiring installations are regulated by the Institution of Engineering and Technology *Requirements for Electrical Installations: IEE Wiring Regulations, BS 7671: 2008*, which are harmonised with IEC 60364. The 17th edition (issued in January 2008) includes new sections for microgeneration and solar photovoltaic systems. The first edition was published in 1882.

In Australia and New Zealand, the AS/NZS 3000 standard, commonly known as the "wiring rules", specifies requirements for the selection and installation of electrical equipment, and the design and testing of such installations. The standard is mandatory in both New Zealand and Australia; therefore, all electrical work covered by the standard must comply.

In European countries, an attempt has been made to harmonise national wiring standards in an IEC standard, IEC 60364 *Electrical Installations for Buildings*. Hence national standards follow an identical system of sections and chapters. However, this standard is not written in such language that it can readily be adopted as a national wiring code. Neither is it designed for field use by electrical tradesmen and inspectors for testing compliance with national wiring standards. By contrast, national codes, such as the NEC or CSA C22.1, generally exemplify the common objectives of IEC 60364, but provide specific rules in a form that allows for guidance of those installing and inspecting electrical systems.

The international standard wire sizes are given in the IEC 60228 standard of the International Electrotechnical Commission. In North America, the American Wire Gauge standard for wire sizes is used.

Colour Code

Colour-coded wires in a flexible plastic electrical conduit found commonly in modern European houses

To enable wires to be easily and safely identified, all common wiring safety codes mandate a colour scheme for the insulation on power conductors. In a typical electrical code, some colour-coding is mandatory, while some may be optional. Many local rules and exceptions exist per country, state or region. Older installations vary in colour codes, and colours may fade with insulation exposure to heat, light and ageing.

As of March 2011, the European Committee for Electrotechnical Standardization (CENELEC) requires the use of green/yellow colour cables as protective conductors, blue as neutral conductors and brown as single-phase conductors. The United States National Electrical Code requires a green or green/yellow protective conductor, a white or grey neutral, and a black single phase.

The United Kingdom requires the use of wire covered with green insulation, to be marked with a prominent yellow stripe, for safe earthing (grounding) connections. This growing international standard was adopted for its distinctive appearance, to reduce the likelihood of dangerous confusion of safety earthing (grounding) wires with other electrical functions, especially by persons affected by red-green colour blindness.

In the UK, phases could be identified as being live by using coloured indicator lights: red, yellow and blue. The new cable colours of brown, black and grey do not lend themselves to coloured indicators. For this reason, three-phase control panels will often use indicator lights of the old colours.

Standard wire insulation colours			
Flexible cable (e.g., extension, power, and lamp cords)			
Region or country	Phases	Neutral	Protective earth/ground
Argentina, Australia, European Union, South Africa (IEC 60446)			
Australia, New Zealand (AS/NZS 3000:2007 3.8.3)	,	,	
Brazil	,		
United States, Canada	□ metallic brass	□metallic silver	, ; ■,□ green/yellow striped
Fixed cable (e.g., in-, on-, or behind-the-wall cables)			
Region or country	Phases	Neutral	Protective earth/ground
Argentina; China; European Union (IEC 60446) from April 2004; the United Kingdom from 31 March 2004 (BS 7671); Hong Kong from July 2007; Singapore from March 2009; Russia since 2009 (GOST R 50462); Ukraine, Belarus, Kazakhstan	, ,		[b]

India, Pakistan; United Kingdom, prior to 31 March 2004 (BS 7671); Hong Kong, prior to 2009; Malaysia and Singapore, prior to February 2011	, ,		[b] • (previously) • no insulation (previously)
Australia, New Zealand (AS/NZS 3000:2007 3.8.1, table 3.4)	• , [c] , , prohibited; any other color permitted • , recommended for single phase • , recommended for multiphase	,[c]	(since about 1980 - Stranded Wire) (since about 1966 - Stranded Wire) Stranded Wire - no insulation; sleeved at the ends (previously)[d]
Brazil	, , ,		
South Africa	• , ; or • ,		[b]

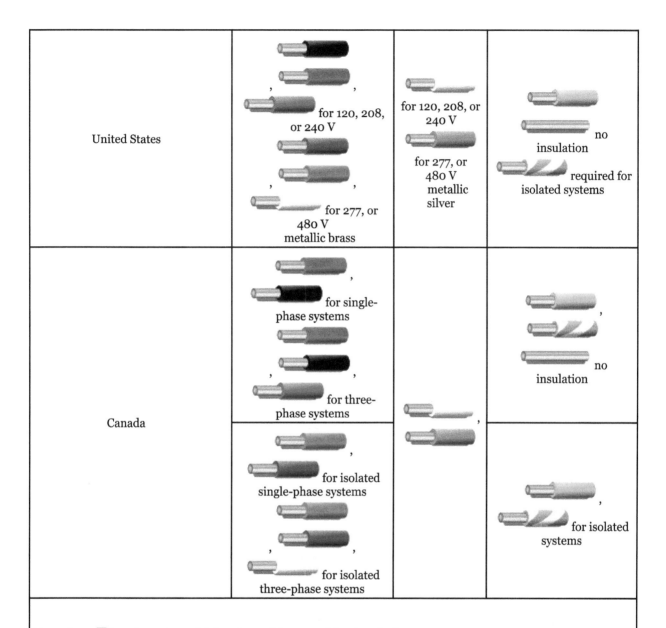

Boxes (e.g., ■ translucent purple) denote markings on wiring terminals.

1. The colours in this table represent the most common and preferred standard colours for wiring; however others may be in use, especially in older installations.

2. Cables may have an uninsulated PE which is sleeved with the appropriate identifying colours at both ends, especially in the UK.

3. Australian and New Zealand wiring standards allow both European and Australian colour codes. Australian-standard phase colours conflict with IEC 60446 colours, where IEC-60446 supported neutral colour (blue) is an allowed phase colour in the Australia/New Zealand standard. Care must be taken when determining system used in existing wiring.

4. The protective earth conductor is now separately insulated throughout all cables.

5. Canadian and American wiring practices are very similar, with ongoing harmonisation efforts.

Wiring Methods

Installing electrical wiring by "chasing" grooves into the masonry structure of the walls of a building

Materials for wiring interior electrical systems in buildings vary depending on:

- Intended use and amount of power demand on the circuit

- Type of occupancy and size of the building

- National and local regulations

- Environment in which the wiring must operate.

Wiring systems in a single family home or duplex, for example, are simple, with relatively low power requirements, infrequent changes to the building structure and layout, usually with dry, moderate temperature and non-corrosive environmental conditions. In a light commercial environment, more frequent wiring changes can be expected, large apparatus may be installed and special conditions of heat or moisture may apply. Heavy industries have more demanding wiring requirements, such as very large currents and higher voltages, frequent changes of equipment layout, corrosive, or wet or explosive atmospheres. In facilities that handle flammable gases or liquids, special rules may govern the installation and wiring of electrical equipment in hazardous areas.

Wires and cables are rated by the circuit voltage, temperature rating and environmental conditions (moisture, sunlight, oil, chemicals) in which they can be used. A wire or cable has a voltage (to neutral) rating and a maximum conductor surface temperature rating. The amount of current a cable or wire can safely carry depends on the installation conditions.

Early Wiring Methods

The first interior power wiring systems used conductors that were bare or covered with cloth, which were secured by staples to the framing of the building or on running boards. Where conductors went through walls, they were protected with cloth tape. Splices were done similarly to telegraph connections, and soldered for security. Underground conductors were insulated with wrappings of cloth tape soaked in pitch, and laid in wooden troughs which were then buried. Such wiring systems were unsatisfactory because of the danger of electrocution and fire, plus the high labour cost for such installations.

Knob and Tube

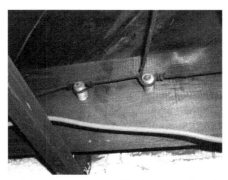

Knob-and-tube wiring (the orange cable is an unrelated extension cord)

The earliest standardised method of wiring in buildings, in common use in North America from about 1880 to the 1930s, was *knob and tube* (K&T) wiring: single conductors were run through cavities between the structural members in walls and ceilings, with ceramic tubes forming protective channels through joists and ceramic knobs attached to the structural members to provide air between the wire and the lumber and to support the wires. Since air was free to circulate over the wires, smaller conductors could be used than required in cables. By arranging wires on opposite sides of building structural members, some protection was afforded against short-circuits that can be caused by driving a nail into both conductors simultaneously.

By the 1940s, the labour cost of installing two conductors rather than one cable resulted in a decline in new knob-and-tube installations. However, the US code still allows new K&T wiring installations in special situations (some rural and industrial applications).

Metal-sheathed Wires

Lead cased electrical wire from a circa 1912 house in Southern England. Two conductors are sheathed in red and black rubber, the central earth wire is bare. These wires are dangerous because the sheath is prone to split if repeatedly flexed.

In the United Kingdom, an early form of insulated cable, introduced in 1896, consisted of two impregnated-paper-insulated conductors in an overall lead sheath. Joints were soldered, and special fittings were used for lamp holders and switches. These cables were similar to underground telegraph and telephone cables of the time. Paper-insulated cables proved unsuitable for interior wiring installations because very careful workmanship was required on the lead sheaths to ensure moisture did not affect the insulation.

A system later invented in the UK in 1908 employed vulcanised-rubber insulated wire enclosed in a strip metal sheath. The metal sheath was bonded to each metal wiring device to ensure earthing continuity.

A system developed in Germany called "Kuhlo wire" used one, two, or three rubber-insulated wires in a brass or lead-coated iron sheet tube, with a crimped seam. The enclosure could also be used as a return conductor. Kuhlo wire could be run exposed on surfaces and painted, or embedded in plaster. Special outlet and junction boxes were made for lamps and switches, made either of porcelain or sheet steel. The crimped seam was not considered as watertight as the *Stannos* wire used in England, which had a soldered sheath.

A somewhat similar system called "concentric wiring" was introduced in the United States around 1905. In this system, an insulated electrical wire was wrapped with copper tape which was then soldered, forming the grounded (return) conductor of the wiring system. The bare metal sheath, at earth potential, was considered safe to touch. While companies such as General Electric manufactured fittings for the system and a few buildings were wired with it, it was never adopted into the US National Electrical Code. Drawbacks of the system were that special fittings were required, and that any defect in the connection of the sheath would result in the sheath becoming energised.

Other Historical Wiring Methods

Other methods of securing wiring that are now obsolete include:

- Re-use of existing gas pipes when converting gas light installations to electric lighting. Insulated conductors were pulled through the pipes that had formerly supplied the gas lamps. Although used occasionally, this method risked insulation damage from sharp edges inside the pipe at each joint.

- Wood mouldings with grooves cut for single conductor wires, covered by a wooden cap strip. These were prohibited in North American electrical codes by 1928. Wooden moulding was also used to some degree in England, but was never permitted by German and Austrian rules.

- A system of flexible twin cords supported by glass or porcelain buttons was used near the turn of the 20th century in Europe, but was soon replaced by other methods.

- During the first years of the 20th century, various patented forms of wiring system such as Bergman and Peschel tubing were used to protect wiring; these used very thin fiber tubes, or metal tubes which were also used as return conductors.

- In Austria, wires were concealed by embedding a rubber tube in a groove in the wall, plastering over it, then removing the tube and pulling wires through the cavity.

Metal moulding systems, with a flattened oval section consisting of a base strip and a snap-on cap channel, were more costly than open wiring or wooden moulding, but could be easily run on wall surfaces. Similar surface mounted raceway wiring systems are still available today.

Cables

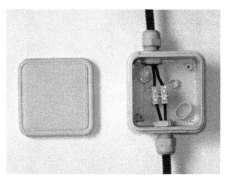

Wiring for extremely wet conditions

Armoured cables with two rubber-insulated conductors in a flexible metal sheath were used as early as 1906, and were considered at the time a better method than open knob-and-tube wiring, although much more expensive.

The first rubber-insulated cables for building wiring were introduced in 1922 with US patent 1458803, Burley, Harry & Rooney, Henry, "Insulated electric wire", issued 1923-06-12, assigned to Boston Insulated Wire And Cable . These were two or more solid copper electrical wires with rubber insulation, plus woven cotton cloth over each conductor for protection of the insulation, with an overall woven jacket, usually impregnated with tar as a protection from moisture. Waxed paper was used as a filler and separator.

Over time, rubber-insulated cables become brittle because of exposure to atmospheric oxygen, so they must be handled with care and are usually replaced during renovations. When switches, socket outlets or light fixtures are replaced, the mere act of tightening connections may cause hardened insulation to flake off the conductors. Rubber insulation further inside the cable often is in better condition than the insulation exposed at connections, due to reduced exposure to oxygen.

The sulphur in vulcanised rubber insulation attacked bare copper wire so the conductors were tinned to prevent this. The conductors reverted to being bare when rubber ceased to be used.

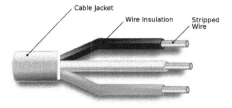

Diagram of a simple electrical cable with three insulated conductors

About 1950, PVC insulation and jackets were introduced, especially for residential wiring. About the same time, single conductors with a thinner PVC insulation and a thin nylon jacket (e.g. US Type THN, THHN, etc.) became common.

The simplest form of cable has two insulated conductors twisted together to form a unit. Such un-jacketed cables with two (or more) conductors are used only for extra low voltage signal and control applications such as doorbell wiring.

US single-phase residential power distribution transformer, showing the two insulated "Line" conductors and the bare "Neutral" conductor (derived from the earthed center-tap of the transformer). The distribution supporting centenaries are also shown.

In North American practice, an overhead cable from a transformer on a power pole to a residential electrical service usually consists of three twisted (triplexed) conductors, with one being a bare protective neutral/earth/ground conductor (which may be made of copper), with the other two being the insulated conductors for both of the two 180 degree out of phase 120 V line voltages normally supplied. However, the earthed/grounded conductor is often a catenary cable (made of steel wire), which is used to support the insulated Line conductors. For additional safety, the ground conductor may be formed into a stranded co-axial layer completely surrounding the phase/line conductors, so that the outermost conductor is grounded.

Copper Conductors

Electrical devices often contain copper conductors because of their multiple beneficial properties, including their high electrical conductivity, tensile strength, ductility, creep resistance, corrosion resistance, thermal conductivity, coefficient of thermal expansion, solderability, resistance to electrical overloads, compatibility with electrical insulators and ease of installation.

Despite competition from other materials, copper remains the preferred electrical conductor in nearly all categories of electrical wiring. For example, copper is used to conduct electricity in high, medium and low voltage power networks, including power generation, power transmission, power distribution, telecommunications, electronics circuitry, data processing, instrumentation, appliances, entertainment systems, motors, transformers, heavy industrial machinery and countless other types of electrical equipment.

Aluminium Conductors

Aluminium wire was common in North American residential wiring from the 1960s to mid-1970s due to the rising cost of copper. Because of its greater resistivity, aluminium wiring requires larger conductors than copper. For instance, instead of 14 AWG (American wire gauge) copper wire, aluminium wiring would need to be 12 AWG on a typical 15 ampere lighting circuit, though local building codes vary.

Terminal blocks for joining aluminium and copper conductors. The terminal blocks may be mounted on a DIN rail.

Solid aluminum conductors were originally made in the 1960's from a utility grade aluminum alloy that had undesirable properties for a building wire, and were indiscriminately used with wiring devices intended for copper conductors and often poorly installed. These practices were found to cause defective connections and potential fire hazards. In the early-1970's new aluminum wire made from one of several special alloys was introduced, and all devices — breakers, switches, receptacles, splice connectors, wire nuts, etc. — were specially designed for the purpose. These newer aluminum wires and special designs address problems with junctions between dissimilar metals, oxidation on metal surfaces and mechanical effects that occur as different metals expand at different rates with increases in temperature.

Unlike copper, aluminium has a tendency to creep or cold-flow under pressure, so older plain steel screw clamped connections could become loose over time. This was be mitigated by using proper installation practices with newer electrical devices (e.g. proper torque on special termination screws), special spring-loaded connectors that apply constant pressure, applying high pressure cold joints in splices and termination fittings, or using a bolted mechanical type clamp wire connector and tightening it to a specified torque.

Also unlike copper, aluminium forms an insulating oxide layer on the surface. This is sometimes addressed by coating aluminium conductors with an antioxidant paste (containing zinc dust in a low-residue polybutene base) at joints, or by applying a mechanical termination designed to break through the oxide layer during installation.

Because of improper design and installation, some junctions to wiring devices would overheat under heavy current load, and cause fires. Revised standards for wire materials and wiring devices (such as the CO/ALR "copper-aluminium-revised" designation) were developed to reduce these problems. Nonetheless, aluminium wiring for residential use has acquired a poor reputation and has fallen out of favour.

Aluminium conductors are still used for bulk power distribution and large feeder circuits, because they cost less than copper wiring, and weigh less, especially in the large sizes needed for heavy current loads. Aluminium conductors must be installed with compatible connectors.

Modern Wiring Materials

Modern non-metallic sheathed cables, such as (US and Canadian) Types NMB and NMC, consist of two to four wires covered with thermoplastic insulation, plus a bare wire for grounding (bonding), surrounded by a flexible plastic jacket. Some versions wrap the individual conductors in paper before the plastic jacket is applied.

Special versions of non-metallic sheathed cables, such as US Type UF, are designed for direct underground burial (often with separate mechanical protection) or exterior use where exposure to ultraviolet radiation (UV) is a possibility. These cables differ in having a moisture-resistant construction, lacking paper or other absorbent fillers, and being formulated for UV resistance.

Rubber-like synthetic polymer insulation is used in industrial cables and power cables installed underground because of its superior moisture resistance.

Insulated cables are rated by their allowable operating voltage and their maximum operating temperature at the conductor surface. A cable may carry multiple usage ratings for applications, for example, one rating for dry installations and another when exposed to moisture or oil.

Generally, single conductor building wire in small sizes is solid wire, since the wiring is not required to be very flexible. Building wire conductors larger than 10 AWG (or about 6 mm²) are stranded for flexibility during installation, but are not sufficiently pliable to use as appliance cord.

Cables for industrial, commercial and apartment buildings may contain many insulated conductors in an overall jacket, with helical tape steel or aluminium armour, or steel wire armour, and perhaps as well an overall PVC or lead jacket for protection from moisture and physical damage. Cables intended for very flexible service or in marine applications may be protected by woven bronze wires. Power or communications cables (e.g., computer networking) that are routed in or through air-handling spaces (plenums) of office buildings are required under the model building code to be either encased in metal conduit, or rated for low flame and smoke production.

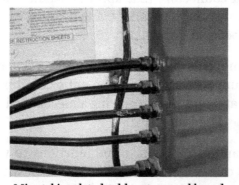

Mineral insulated cables at a panel board

For some industrial uses in steel mills and similar hot environments, no organic material gives satisfactory service. Cables insulated with compressed mica flakes are sometimes used. Another form of high-temperature cable is a mineral insulated cable, with individual conductors placed within a copper tube and the space filled with magnesium oxide powder. The whole assembly is drawn down to smaller sizes, thereby compressing the powder. Such cables have a certified fire resistance rating and are more costly than non-fire rated cable. They have little flexibility and behave more like rigid conduit rather than flexible cables.

Because multiple conductors bundled in a cable cannot dissipate heat as easily as single insulated conductors, those circuits are always rated at a lower "ampacity". Tables in electrical safety codes give the maximum allowable current for a particular size of conductor, for the voltage and temperature rating at the surface of the conductor for a given physical environment, including the insulation type and thickness. The allowable current will be different for wet or dry, for hot (attic) or cool (underground) locations. In a run of cable through several areas, the most severe area will determine the appropriate rating of the overall run.

Cables usually are secured by special fittings where they enter electrical apparatus; this may be a simple screw clamp for jacketed cables in a dry location, or a polymer-gasketed cable connector that mechanically engages the armour of an armoured cable and provides a water-resistant connection. Special cable fittings may be applied to prevent explosive gases from flowing in the interior of jacketed cables, where the cable passes through areas where inflammable gases are present. To prevent loosening of the connections of individual conductors of a cable, cables must be supported near their entrance to devices and at regular intervals through their length. In tall buildings, special designs are required to support the conductors of vertical runs of cable. Usually, only one cable per fitting is allowed unless the fitting is otherwise rated.

Special cable constructions and termination techniques are required for cables installed in ocean-going vessels; in addition to electrical safety and fire safety, such cables may also be required to be pressure-resistant where they penetrate bulkheads of a ship. Resistance to corrosion caused by salt water or salt spray is also required.

Raceways

Electrical conduit risers, seen inside fire-resistance rated shaft, as seen entering bottom of a firestop. The firestop is made of firestop mortar on top, rockwool on the bottom. Raceways are used to protect cables from damage.

Insulated wires may be run in one of several forms of a raceway between electrical devices. This may be a specialised bendable pipe, called a conduit, or one of several varieties of metal (rigid steel or aluminium) or non-metallic (PVC or HDPE) tubing. Rectangular cross-section metal or PVC wire troughs (North America) or trunking (UK) may be used if many circuits are required. Wires run underground may be run in plastic tubing encased in concrete, but metal elbows may be used in severe pulls. Wiring in exposed areas, for example factory floors, may be run in cable trays or rectangular raceways having lids.

Where wiring, or raceways that hold the wiring, must traverse fire-resistance rated walls and floors, the openings are required by local building codes to be firestopped. In cases where safety-critical wiring must be kept operational during an accidental fire, fireproofing must be applied to maintain circuit integrity in a manner to comply with a product's certification listing. The nature and thickness of any passive fire protection materials used in conjunction with wiring and raceways has a quantifiable impact upon the ampacity derating, because the thermal insulation properties needed for fire resistance also inhibit air cooling of power conductors.

A cable tray can be used in stores and dwellings

Cable trays are used in industrial areas where many insulated cables are run together. Individual cables can exit the tray at any point, simplifying the wiring installation and reducing the labour cost for installing new cables. Power cables may have fittings in the tray to maintain clearance between the conductors, but small control wiring is often installed without any intentional spacing between cables.

Note that cable trays are very common in two-way radio sites where they are used for antenna cables, some of which can be 3 1/8 inches (8 cm) or even larger. Local codes may preclude mixing power cables with antenna cables in the same tray.

Since wires run in conduits or underground cannot dissipate heat as easily as in open air, and since adjacent circuits contribute induced currents, wiring regulations give rules to establish the current capacity (ampacity).

Special sealed fittings are used for wiring routed through potentially explosive atmospheres.

Bus Bars, Bus Duct, Cable Bus

Topside of firestop with penetrants consisting of electrical conduit on the left and a bus duct on the right. The firestop consists of firestop mortar on top and rockwool on the bottom, for a 2-hour fire-resistance rating.

For very high currents in electrical apparatus, and for high currents distributed through a building, bus bars can be used. (The term "bus" is a contraction of the Latin *omnibus* – meaning "for all".) Each live conductor of such a system is a rigid piece of copper or aluminium, usually in flat bars (but sometimes as tubing or other shapes). Open bus bars are never used in publicly accessible areas, although they are used in manufacturing plants and power company switch yards to gain the benefit of air cooling. A variation is to use heavy cables, especially where it is desirable to transpose or "roll" phases.

In industrial applications, conductor bars are often pre-assembled with insulators in grounded enclosures. This assembly, known as bus duct or busway, can be used for connections to large switchgear or for bringing the main power feed into a building. A form of bus duct known as "plug-in bus" is used to distribute power down the length of a building; it is constructed to allow tap-off switches or motor controllers to be installed at designated places along the bus. The big advantage of this scheme is the ability to remove or add a branch circuit without removing voltage from the whole duct.

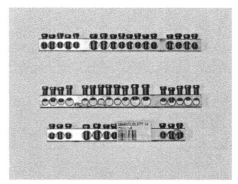

Busbars for distributing PE (ground)

Bus ducts may have all phase conductors in the same enclosure (non-isolated bus), or may have each conductor separated by a grounded barrier from the adjacent phases (segregated bus). For conducting large currents between devices, a cable bus is used.

For very large currents in generating stations or substations, where it is difficult to provide circuit protection, an isolated-phase bus is used. Each phase of the circuit is run in a separate grounded metal enclosure. The only fault possible is a phase-to-ground fault, since the enclosures are separated. This type of bus can be rated up to 50,000 amperes and up to hundreds of kilovolts (during normal service, not just for faults), but is not used for building wiring in the conventional sense.

Electrical Panels

Electrical panels in an electrical service room at St. Mary's
Pulp and Paper, Sault Ste. Marie, Ontario, Canada, April
1996

Electrical panels, cables and firestops in an electrical service room at a paper mill in Ontario, Canada

Electrical panels are easily accessible junction boxes used to reroute and switch electrical services. The term is often used to refer to circuit breaker panels or fuseboxes. Local codes can specify physical clearance around the panels.

Degradation by Pests

Rasberry crazy ants have been known to consume the insides of electrical wiring installations, preferring DC over AC currents. This behaviour is not well understood by scientists.

Squirrels, rats and other rodents may gnaw on unprotected wiring, causing fire and shock hazards. This is especially true of PVC-insulated telephone and computer network cables. Several techniques have been developed to deter these pests, including insulation loaded with pepper dust.

Daisy Chain (Electrical Engineering)

A series of devices connected in a daisy chain layout

In electrical and electronic engineering a daisy chain is a wiring scheme in which multiple devices are wired together in sequence or in a ring. Other than a full, single loop, systems which contain internal loops cannot be called daisy chains.

Daisy chains may be used for power, analog signals, digital data, or a combination thereof.

The term daisy chain may refer either to large scale devices connected in series, such as a series of power strips plugged into each other to form a single long line of strips, or to the wiring patterns embedded inside of devices. Other examples of devices which can be used to form daisy chains are those based on USB, FireWire, Thunderbolt and Ethernet cables.

Signal Transmission

For analog signals, connections usually comprise a simple electrical bus and, especially in the case of a chain of many devices, may require the use of one or more repeaters or amplifiers within the chain to counteract attenuation (the natural loss of energy in such a system). Digital signals between devices may also comprise a simple electrical bus, in which case a bus terminator may be needed on the last device in the chain. However, unlike analog signals, because digital signals are discrete, they may also be electrically regenerated, but not modified, by any device in the chain.

Types

Computer Hardware

Some hardware can be attached to a computing system in a daisy chain configuration by connecting each component to another similar component, rather than directly to the computing system that uses the component. Only the last component in the chain directly connects to the computing system. For example, chaining multiple components that each have a UART port to each other. The components must also behave cooperatively. e.g., only one seizes the communications bus at a time.

- SCSI is an example of a digital system that is electrically a bus, in the case of external devices, is physically wired as a daisy chain. Since the network is electrically a bus, it must be terminated and this may be done either by plugging a terminator into the last device or selecting an option to make the device terminate internally.

- MIDI devices are usually designed to be wired in a daisy chain. It is normal for a device to have both a THRU port and an OUT port and often both can be used for chaining. The THRU port transmits the information through with minimal delay and no alteration, while the OUT port sends a completely regenerated signal and may add, remove, or change messages, at the cost of some delay in doing so. The difference can result in the signals arriving at different times; if the chain is long enough, it will be distorted so much that the system can become unreliable or non-functional.

- Some Serial Peripheral Interface Bus (SPI) IC products are designed with daisy chain capability.

- All JTAG integrated circuits should support daisy chaining according to JTAG daisy chaining guidelines.

- Thunderbolt (interface) also supports daisy-chained devices such as RAID arrays and computer monitors.

- The Hexbus is the 10-wire bus of Texas Instruments, used in the TI-99/4A, CC-40 and TI-74.

Network Topology

Any particular daisy chain forms one of two network topologies:

- Linear topology: For example, A-B-C-D-E, A-B-C-D-E & C-M-N-O (branched at C) are daisy chain.

- Ring topology: there is a loop connection back from the last device to the first. For example, A-B-C-D-E-A (loop). This is often called a "daisy chain loop".

System Access

Users can daisy chain computing sessions together. Using services such as Telnet or SSH, the user creates a session on a second computer via Telnet, and from the second session, Telnets to a third and so on. Another typical example is the "terminal session inside a terminal session" using RDP. Reasons to create daisy chains include connecting to a system on a non-routed network via a gateway system, preserving sessions on the initial computer while working on a second computer, to save bandwidth or improve connectivity on an unstable network by first connecting to a better connected machine. A less wholesome purpose is camouflaging activity while engaged in cybercrime.

Signal (Electrical Engineering)

A signal as referred to in communication systems, signal processing, and electrical engineering is a function that "conveys information about the behavior or attributes of some phenomenon". In the physical world, any quantity exhibiting variation in time or variation in space (such as an image) is potentially a signal that might provide information on the status of a physical system, or convey a message between observers, among other possibilities. The *IEEE Transactions on Signal Processing* states that the term "signal" includes audio, video, speech, image, communication, geophysical, sonar, radar, medical and musical signals.

Typically, signals are provided by a sensor, and often the original form of a signal is converted to another form of energy using a transducer. For example, a microphone converts an acoustic signal to a voltage waveform, and a speaker does the reverse.

The formal study of the information content of signals is the field of information theory. The information in a signal is usually accompanied by noise. The term *noise* usually means an undesirable random disturbance, but is often extended to include unwanted signals conflicting with the desired signal (such as crosstalk). The prevention of noise is covered in part under the heading of signal integrity. The separation of desired signals from a background is the field of signal recovery, one branch of which is estimation theory, a probabilistic approach to suppressing random disturbances.

Engineering disciplines such as electrical engineering have led the way in the design, study, and implementation of systems involving transmission, storage, and manipulation of information. In the latter half of the 20th century, electrical engineering itself separated into several disciplines, specialising in the design and analysis of systems that manipulate physical signals; electronic engineering and computer engineering as examples; while design engineering developed to deal with functional design of man–machine interfaces.

Definitions

Definitions specific to sub-fields are common. For example, in information theory, a *signal* is a codified message, that is, the sequence of states in a communication channel that encodes a message.

In the context of signal processing, arbitrary binary data streams are not considered as signals, but only analog and digital signals that are representations of analog physical quantities.

In a *communication system*, a *transmitter* encodes a *message* to a signal, which is carried to a *receiver* by the communications *channel*. For example, the words "Mary had a little lamb" might be the message spoken into a telephone. The telephone transmitter converts the sounds into an electrical voltage signal. The signal is transmitted to the receiving telephone by wires; at the receiver it is reconverted into sounds.

In telephone networks, signalling, for example common-channel signaling, refers to phone number and other digital control information rather than the actual voice signal.

Signals can be categorized in various ways. The most common distinction is between discrete and continuous spaces that the functions are defined over, for example discrete and continuous time domains. Discrete-time signals are often referred to as *time series* in other fields. Continuous-time signals are often referred to as *continuous signals* even when the signal functions are not continuous; an example is a square-wave signal.

A second important distinction is between discrete-valued and continuous-valued. Particularly in digital signal processing a digital signal is sometimes defined as a sequence of discrete values, that may or may not be derived from an underlying continuous-valued physical process. In other contexts, digital signals are defined as the continuous-time waveform signals in a digital system, representing a bit-stream. In the first case, a signal that is generated by means of a digital modulation method is considered as converted to an analog signal, while it is considered as a digital signal in the second case.

Another important property of a signal (actually, of a statistically defined class of signals) is its entropy or *information content.*

Analog and Digital Signals

A digital signal has two or more distinguishable waveforms, in this example, high voltage and low voltages, each of which can be mapped onto a digit. Characteristically, noise can be removed from digital signals provided it is not too large.

Two main types of signals encountered in practice are *analog* and *digital*. The figure shows a digital signal that results from approximating an analog signal by its values at particular time instants. Digital signals are *quantized*, while analog signals are continuous.

Digital signals often arise via sampling of analog signals, for example, a continually fluctuating voltage on a line that can be digitized by an analog-to-digital converter circuit, wherein the circuit will read the voltage level on the line, say, every 50 microseconds and employing a fixed number of bits. The resulting stream of numbers is stored as digital data on a discrete-time and quantized-amplitude signal. Computers and other digital devices are restricted to discrete time.

Time Discretization

One of the fundamental distinctions between different types of signals is between continuous and discrete time. In the mathematical abstraction, the domain of a continuous-time (CT) signal is the set of real numbers (or some interval thereof), whereas the domain of a discrete-time (DT) signal is the set of integers (or some interval). What these integers represent depends on the nature of the signal; most often it is time.

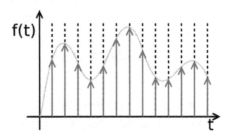

Discrete-time signal created from a continuous signal by sampling

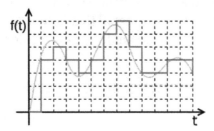

Digital signal resulting from approximation to an analog signal, which is a continuous function of time

If for a signal, the quantities are defined only on a discrete set of times, we call it a discrete-time signal. A simple source for a discrete time signal is the sampling of a continuous signal, approximating the signal by a sequence of its values at particular time instants.

A discrete-time real (or complex) signal can be seen as a function from (a subset of) the set of integers (the index labeling time instants) to the set of real (or complex) numbers (the function values at those instants).

A continuous-time real (or complex) signal is any real-valued (or complex-valued) function which is defined at every time t in an interval, most commonly an infinite interval.

Amplitude Quantization

If a signal is to be represented as a sequence of numbers, it is impossible to maintain exact precision - each number in the sequence must have a finite number of digits. As a result, the values of such a signal belong to a finite set; in other words, it is quantized. Quantization is the process of converting a continuous analog audio signal to a digital signal with discrete numerical values.

Examples of Signals

Signals in nature can be converted to electronic signals by various sensors. Some examples are:

- *Motion.* The motion of an object can be considered to be a signal, and can be monitored by various sensors to provide electrical signals. For example, radar can provide an electromagnetic signal for following aircraft motion. A motion signal is one-dimensional (time), and the range is generally three-dimensional. Position is thus a 3-vector signal; position and orientation of a rigid body is a 6-vector signal. Orientation signals can be generated using a gyroscope.

- *Sound.* Since a sound is a vibration of a medium (such as air), a sound signal associates a pressure value to every value of time and three space coordinates. A sound signal is converted to an electrical signal by a microphone, generating a voltage signal as an analog of the sound signal, making the sound signal available for further signal processing. Sound signals can be sampled at a discrete set of time points; for example, compact discs (CDs) contain discrete signals representing sound, recorded at 44,100 samples per second; each sample contains data for a left and right channel, which may be considered to be a 2-vector signal (since CDs are recorded in stereo). The CD encoding is converted to an electrical signal by reading the information with a laser, converting the sound signal to an optical signal.

- *Images.* A picture or image consists of a brightness or color signal, a function of a two-dimensional location. The object's appearance is presented as an emitted or reflected electromagnetic wave, one form of electronic signal. It can be converted to voltage or current waveforms using devices such as the charge-coupled device. A 2D image can have a continuous spatial domain, as in a traditional photograph or painting; or the image can be discretized in space, as in a raster scanned digital image. Color images are typically represented as a combination of images in three primary colors, so that the signal is vector-valued with dimension three.

- *Videos.* A video signal is a sequence of images. A point in a video is identified by its two-dimensional position and by the time at which it occurs, so a video signal has a three-dimensional domain. Analog video has one continuous domain dimension (across a scan line) and two discrete dimensions (frame and line).

- Biological *membrane potentials.* The value of the signal is an electric potential ("voltage"). The domain is more difficult to establish. Some cells or organelles have the same membrane potential throughout; neurons generally have different potentials at different points. These signals have very low energies, but are enough to make nervous systems work; they can be measured in aggregate by the techniques of electrophysiology.

Other examples of signals are the output of a thermocouple, which conveys temperature information, and the output of a pH meter which conveys acidity information.

Signal Processing

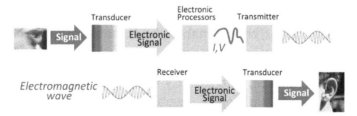

Signal transmission using electronic signals

A typical role for signals is in signal processing. A common example is signal transmission between different locations. The embodiment of a signal in electrical form is made by a transducer that converts the signal from its original form to a waveform expressed as a current (I) or a voltage (V), or an electromagnetic waveform, for example, an optical signal or radio transmission. Once expressed as an electronic signal, the signal is available for further processing by electrical devices such as electronic amplifiers and electronic filters, and can be transmitted to a remote location by electronic transmitters and received using electronic receivers.

Signals and Systems

In Electrical engineering programs, a class and field of study known as "signals and systems" (S and S) is often seen as the "cut class" for EE careers, and is dreaded by some students as such. Depending on the school, undergraduate EE students generally take the class as juniors or seniors, normally depending on the number and level of previous linear algebra and differential equation classes they have taken.

The field studies input and output signals, and the mathematical representations between them known as systems, in four domains: Time, Frequency, s and z. Since signals and systems are both studied in these four domains, there are 8 major divisions of study. As an example, when working with continuous time signals (t), one might transform from the time domain to a frequency or s domain; or from discrete time (n) to frequency or z domains. Systems also can be transformed between these domains like signals, with continuous to s and discrete to z.

Although S and S falls under and includes all the topics covered in this article, as well as Analog signal processing and Digital signal processing, it actually is a subset of the field of Mathematical modeling. The field goes back to RF over a century ago, when it was all analog, and generally continuous. Today, software has taken the place of much of the analog circuitry design and analysis, and even continuous signals are now generally processed digitally. Ironically, digital signals also are processed continuously in a sense, with the software doing calculations between discrete signal "rests" to prepare for the next input/transform/output event.

In past EE curricula S and S, as it is often called, involved circuit analysis and design via mathematical modeling and some numerical methods, and was updated several decades ago with Dynamical systems tools including differential equations, and recently, Lagrangians. The difficulty of the field at that time included the fact that not only mathematical modeling, circuits, signals and complex systems were being modeled, but physics as well, and a deep knowledge of electrical (and now electronic) topics also was involved and required.

Today, the field has become even more daunting and complex with the addition of circuit, systems and signal analysis and design languages and software, from MATLAB and Simulink to NumPy, VHDL, PSpice, Verilog and even Assembly language. Students are expected to understand the tools as well as the mathematics, physics, circuit analysis, and transformations between the 8 domains.

Because mechanical engineering topics like friction, dampening etc. have very close analogies in signal science (inductance, resistance, voltage, etc.), many of the tools originally used in ME transformations (Laplace and Fourier transforms, Lagrangians, sampling theory, probability, difference equations, etc.) have now been applied to signals, circuits, systems and their components, analysis and design in EE. Dynamical systems that involve noise, filtering and other random or chaotic attractors and repellors have now placed stochastic sciences and statistics between the more deterministic discrete and continuous functions in the field. (Deterministic as used here means signals that are completely determined as functions of time).

EE taxonomists are still not decided where S&S falls within the whole field of signal processing vs. circuit analysis and mathematical modeling, but the common link of the topics that are covered in the course of study has brightened boundaries with dozens of books, journals, etc. called Signals and Systems, and used as text and test prep for the EE, as well as, recently, computer engineering exams. The Hsu general reference given below is a good example, with a new edition scheduled for late 2013/ early 2014.

References

- C22.1-15—Canadian Electrical Code, Part I: Safety Standard for Electrical Installations (23rd ed.). Canadian Standards Association. 2015. Rules 4-038, 24-208(c). ISBN 978-1-77139-718-6.

- Robert M. Black, The History of Electric Wires and Cable, Peter Pergrinus Ltd. London, 1983 ISBN 0-86341-001-4, pp. 155–158

- Joseph, Günter, 1999, Copper: Its Trade, Manufacture, Use, and Environmental Status, Kundig, Konrad J.A. (ed.), ASM International, ISBN 0871706563, pp. 141–192, 331–375

- T. H. Wilmshurst (1990). Signal Recovery from Noise in Electronic Instrumentation (2nd ed.). CRC Press. pp. 11 ff. ISBN 0750300582.

- The optical reading process is described by Mark L. Chambers (2004). CD & DVD Recording for Dummies (2nd ed.). John Wiley & Sons. p. 13. ISBN 0764559567.

- M.J. Roberts (2011). Signals and Systems: Analysis Using Transform Methods & MATLAB. New York: McGraw Hill. ISBN 978-0073380681.

- "Generating Power to Your House - How Power Grids Work - HowStuffWorks". HowStuffWorks. Retrieved 21 February 2016.

- Noel Williams, Jeffrey S. Sargen. "NEC Q and A: Questions and Answers on the National Electrical Code". p. 117. Retrieved 4 January 2016.

Physical Laws Related to Electrical Engineering

5

The physical laws related to electrical engineering are discussed in this chapter. Some of these laws are Ampère's circuital law, Kirchhoff's circuit laws, Coulomb's law and Ohm's law. This chapter helps the readers in understanding the laws related to electrical engineering.

Ampère's Circuital Law

In classical electromagnetism, Ampère's circuital law relates the integrated magnetic field around a closed loop to the electric current passing through the loop. James Clerk Maxwell (not Ampère) derived it using hydrodynamics in his 1861 paper *On Physical Lines of Force* and it is now one of the Maxwell equations, which form the basis of classical electromagnetism.

Maxwell's Original Circuital Law

The original form of Maxwell's circuital law, which he derived in his 1855 paper *On Faraday's Lines of Force* based on an analogy to hydrodynamics, relates magnetic fields to electric currents that produce them. It determines the magnetic field associated with a given current, or the current associated with a given magnetic field.

The original circuital law is only a correct law of physics in a magnetostatic situation, where the system is static except possibly for continuous steady currents within closed loops. For systems with electric fields that change over time, the original law must be modified to include a term known as Maxwell's correction.

Equivalent Forms

The original circuital law can be written in several different forms, which are all ultimately equivalent:

- An "integral form" and a "differential form". The forms are exactly equivalent, and related by the Kelvin–Stokes theorem.

- Forms using SI units, and those using cgs units. Other units are possible, but rare.

- Forms using either B or H magnetic fields. These two forms use the total current density and free current density, respectively. The B and H fields are related by the constitutive equation: $\mathbf{B} = \mu_0 \mathbf{H}$ where μ_0 is the magnetic constant.

Explanation

The integral form of the original circuital law is a line integral of the magnetic field around some closed curve C (arbitrary but must be closed). The curve C in turn bounds both a surface S which the electric current passes through (again arbitrary but not closed—since no three-dimensional volume is enclosed by S), and encloses the current. The mathematical statement of the law is a relation between the total amount of magnetic field around some path (line integral) due to the current which passes through that enclosed path (surface integral).

In terms of total current, (which is the sum of both free current and bound current) the line integral of the magnetic B-field (in tesla, T) around closed curve C is proportional to the total current I_{enc} passing through a surface S (enclosed by C). In terms of free current, the line integral of the magnetic H-field (in ampere per metre, Am^{-1}) around closed curve C equals the free current $I_{f,\,enc}$ through a surface S.

Forms of the original circuital law written in SI units		
	Integral form	**Differential form**
Using B-field and total current	$\oint_C \mathbf{B} \cdot d\ell = \mu_0 \iint_S \mathbf{J} \cdot d\mathbf{S} = \mu_0 I_{enc}$	$\nabla \times \mathbf{B} = \mu_0 \mathbf{J}$
Using H-field and free current	$\oint_C \mathbf{H} \cdot d\ell = \iint_S \mathbf{J}_f \cdot d\mathbf{S} = I_{f,\,enc}$	$\nabla \times \mathbf{H} = \mathbf{J}_f$

- J is the total current density (in ampere per square metre, Am^{-2}).
- J_f is the free current density only.
- $\oint_c$ is the closed line integral around the closed curve C,
- $\iint$ denotes a 2d surface integral over S enclosed by C
- $\cdot$ is the vector dot product,
- $d\ell$ is an infinitesimal element (a differential) of the curve C (i.e. a vector with magnitude equal to the length of the infinitesimal line element, and direction given by the tangent to the curve C)
- dS is the vector area of an infinitesimal element of surface S (that is, a vector with magnitude equal to the area of the infinitesimal surface element, and direction normal to surface S. The direction of the normal must correspond with the orientation of C by the right hand rule).
- $\nabla \times$ is the curl operator.

Ambiguities and Sign Conventions

There are a number of ambiguities in the above definitions that require clarification and a choice of convention.

1. First, three of these terms are associated with sign ambiguities: the line integral $\oint_C$ could go around the loop in either direction (clockwise or counterclockwise); the vector area dS could point in either of the two directions normal to the surface; and I_{enc} is the net current passing through the surface S, meaning the current passing through in one direction, minus the current in the other direction—but either direction could be chosen as positive. These ambiguities are resolved by the right-hand rule: With the palm of the right-hand toward the area of integration, and the index-finger pointing along the direction of line-integration, the outstretched thumb points in the direction that must be chosen for the vector area dS. Also the current passing in the same direction as dS must be counted as positive. The right hand grip rule can also be used to determine the signs.

2. Second, there are infinitely many possible surfaces S that have the curve C as their border. (Imagine a soap film on a wire loop, which can be deformed by moving the wire). Which of those surfaces is to be chosen? If the loop does not lie in a single plane, for example, there is no one obvious choice. The answer is that it does not matter; it can be proven that any surface with boundary C can be chosen.

Free Current Versus Bound Current

The electric current that arises in the simplest textbook situations would be classified as "free current"—for example, the current that passes through a wire or battery. In contrast, "bound current" arises in the context of bulk materials that can be magnetized and/or polarized. (All materials can to some extent.)

When a material is magnetized (for example, by placing it in an external magnetic field), the electrons remain bound to their respective atoms, but behave as if they were orbiting the nucleus in a particular direction, creating a microscopic current. When the currents from all these atoms are put together, they create the same effect as a macroscopic current, circulating perpetually around the magnetized object. This magnetization current J_M is one contribution to "bound current".

The other source of bound current is bound charge. When an electric field is applied, the positive and negative bound charges can separate over atomic distances in polarizable materials, and when the bound charges move, the polarization changes, creating another contribution to the "bound current", the polarization current J_P.

The total current density J due to free and bound charges is then:

$$\mathbf{J} = \mathbf{J}_f + \mathbf{J}_M + \mathbf{J}_P$$

with J_f the "free" or "conduction" current density.

All current is fundamentally the same, microscopically. Nevertheless, there are often practical reasons for wanting to treat bound current differently from free current. For example, the bound current usually originates over atomic dimensions, and one may wish to take advantage of a simpler theory intended for larger dimensions. The result is that the more microscopic Ampère's circuital law, expressed in terms of B and the microscopic current (which includes free, magnetization and polarization currents), is sometimes put into the equivalent form below in terms of H and the free current only. For a detailed definition of free current and bound current, and the proof that the two formulations are equivalent.

Shortcomings of the Original Formulation of the Circuital Law

There are two important issues regarding the circuital law that require closer scrutiny. First, there is an issue regarding the continuity equation for electrical charge. In vector calculus, the identity for the divergence of a curl states that a vector field's curl divergence must always be zero. Hence

$$\nabla \cdot (\nabla \times \mathbf{B}) = 0$$

and so the original Ampère's circuital law implies that

$$\nabla \cdot \mathbf{J} = 0.$$

But in general, reality follows the continuity equation for electric charge:

$$\nabla \cdot \mathbf{J} = -\frac{\partial \rho}{\partial t}$$

which is non-zero for a time-varying charge density. An example occurs in a capacitor circuit where time-varying charge densities exist on the plates.

Second, there is an issue regarding the propagation of electromagnetic waves. For example, in free space, where

$$\mathbf{J} = 0,$$

The circuital law implies that

$$\nabla \times \mathbf{B} = 0$$

but experimental tests actually show that

$$\nabla \times \mathbf{B} = \frac{1}{c^2} \frac{\partial \mathbf{E}}{\partial t}.$$

To treat these situations, the contribution of displacement current must be added to the current term in the circuital law.

James Clerk Maxwell conceived of displacement current as a polarization current in the dielectric vortex sea, which he used to model the magnetic field hydrodynamically and mechanically. He added this displacement current to Ampère's circuital law at equation (112) in his 1861 paper *On Physical Lines of Force*.

Displacement Current

In free space, the displacement current is related to the time rate of change of electric field.

In a dielectric the above contribution to displacement current is present too, but a major contribution to the displacement current is related to the polarization of the individual molecules of the dielectric material. Even though charges cannot flow freely in a dielectric, the charges in molecules can move a little under the influence of an electric field. The positive and negative charges in molecules separate under the applied field, causing an increase in the state of polarization, expressed as the polarization density P. A changing state of polarization is equivalent to a current.

Both contributions to the displacement current are combined by defining the displacement current as:

$$\mathbf{J}_D = \frac{\partial}{\partial t}\mathbf{D}(\mathbf{r}, t),$$

where the electric displacement field is defined as:

$$\mathbf{D} = \varepsilon_0\mathbf{E} + \mathbf{P} = \varepsilon_0\varepsilon_r\mathbf{E},$$

where ε_0 is the electric constant, ε_r the relative static permittivity, and P is the polarization density. Substituting this form for D in the expression for displacement current, it has two components:

$$\mathbf{J}_D = \varepsilon_0\frac{\partial \mathbf{E}}{\partial t} + \frac{\partial \mathbf{P}}{\partial t}.$$

The first term on the right hand side is present everywhere, even in a vacuum. It doesn't involve any actual movement of charge, but it nevertheless has an associated magnetic field, as if it were an actual current. Some authors apply the name *displacement current* to only this contribution.

The second term on the right hand side is the displacement current as originally conceived by Maxwell, associated with the polarization of the individual molecules of the dielectric material.

Maxwell's original explanation for displacement current focused upon the situation that occurs in dielectric media. In the modern post-aether era, the concept has been extended to apply to situations with no material media present, for example, to the vacuum between the plates of a charging vacuum capacitor. The displacement current is justified today because it serves several requirements of an electromagnetic theory: correct prediction of magnetic fields in regions where no free current flows; prediction of wave propagation of electromagnetic fields; and conservation of electric charge in cases where charge density is time-varying.

Extending the Original Law: The Maxwell–Ampère Equation

Next, the circuital equation is extended by including the polarization current, thereby remedying the limited applicability of the original circuital law.

Treating free charges separately from bound charges, The equation including Maxwell's correction in terms of the H-field is:

$$\oint_c \mathbf{H}\cdot d\ell = \iint_s \left(\mathbf{J}_f + \frac{\partial \mathbf{D}}{\partial t}\right)\cdot d\mathbf{S}$$

(integral form), where H is the magnetic H field (also called "auxiliary magnetic field", "magnetic field intensity", or just "magnetic field"), D is the electric displacement field, and J_f is the enclosed conduction current or free current density. In differential form,

$$\nabla \times \mathbf{H} = \mathbf{J}_f + \frac{\partial \mathbf{D}}{\partial t}.$$

On the other hand, treating all charges on the same footing (disregarding whether they are bound or free charges), the generalized Ampère's equation, also called the Maxwell–Ampère equation, is in integral form:

$$\oint_c \mathbf{B} \cdot d\ell = \iint_s \left(\mu_0 \mathbf{J} + \mu_0 \varepsilon_0 \frac{\partial \mathbf{E}}{\partial t} \right) \cdot d\mathbf{S}$$

In differential form,

$$\nabla \times \mathbf{B} = \mu_0 \mathbf{J} + \mu_0 \varepsilon_0 \frac{\partial \mathbf{E}}{\partial t}$$

In both forms J includes magnetization current density as well as conduction and polarization current densities. That is, the current density on the right side of the Ampère–Maxwell equation is:

$$\mathbf{J}_f + \mathbf{J}_D + \mathbf{J}_M = \mathbf{J}_f + \mathbf{J}_P + \mathbf{J}_M + \varepsilon_0 \frac{\partial \mathbf{E}}{\partial t} = \mathbf{J} + \varepsilon_0 \frac{\partial \mathbf{E}}{\partial t},$$

where current density J_D is the *displacement current*, and J is the current density contribution actually due to movement of charges, both free and bound. Because $\nabla \cdot \mathbf{D} = \rho$, the charge continuity issue with Ampère's original formulation is no longer a problem. Because of the term in $\varepsilon_0 \partial \mathbf{E} / \partial t$, wave propagation in free space now is possible.

With the addition of the displacement current, Maxwell was able to hypothesize (correctly) that light was a form of electromagnetic wave. See electromagnetic wave equation for a discussion of this important discovery.

Proof of Equivalence

Ampère's Circuital Law in Cgs Units

In cgs units, the integral form of the equation, including Maxwell's correction, reads

$$\oint_c \mathbf{B} \cdot d\ell = \frac{1}{c} \iint_s \left(4\pi \mathbf{J} + \frac{\partial \mathbf{E}}{\partial t} \right) \cdot d\mathbf{S}$$

where c is the speed of light.

The differential form of the equation (again, including Maxwell's correction) is

$$\nabla \times \mathbf{B} = \frac{1}{c} \left(4\pi \mathbf{J} + \frac{\partial \mathbf{E}}{\partial t} \right).$$

Kirchhoff's Circuit Laws

Kirchhoff's circuit laws are two equalities that deal with the current and potential difference (commonly known as voltage) in the lumped element model of electrical circuits. They were first described in 1845 by German physicist Gustav Kirchhoff. This generalized the work of Georg Ohm and preceded the work of Maxwell. Widely used in electrical engineering, they are also called Kirchhoff's rules or simply Kirchhoff's laws.

Both of Kirchhoff's laws can be understood as corollaries of the Maxwell equations in the low-frequency limit. They are accurate for DC circuits, and for AC circuits at frequencies where the wavelengths of electromagnetic radiation are very large compared to the circuits.

Kirchhoff's Current Law (KCL)

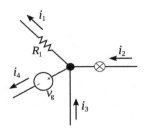

The current entering any junction is equal to the current leaving that junction. $i_2 + i_3 = i_1 + i_4$

This law is also called Kirchhoff's first law, Kirchhoff's point rule, or Kirchhoff's junction rule (or nodal rule).

The principle of conservation of electric charge implies that:

> At any node (junction) in an electrical circuit, the sum of currents flowing into that node is equal to the sum of currents flowing out of that node

or equivalently

> The algebraic sum of currents in a network of conductors meeting at a point is zero.

Recalling that current is a signed (positive or negative) quantity reflecting direction towards or away from a node, this principle can be stated as:

$$\sum_{k=1}^{n} I_k = 0$$

n is the total number of branches with currents flowing towards or away from the node.

This formula is valid for complex currents:

$$\sum_{k=1}^{n} \tilde{I}_k = 0$$

The law is based on the conservation of charge whereby the charge (measured in coulombs) is the product of the current (in amperes) and the time (in seconds).

Uses

A matrix version of Kirchhoff's current law is the basis of most circuit simulation software, such as SPICE. Kirchhoff's current law combined with Ohm's Law is used in nodal analysis.

KCL is applicable to any lumped network irrespective of the nature of the network; whether unilateral or bilateral, active or passive, linear or non-linear.

Kirchhoff's Voltage Law (KVL)

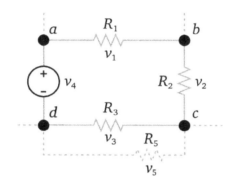

The sum of all the voltages around a loop is equal to zero.

$$V_1 + V_2 + V_3 - V_4 = 0$$

This law is also called Kirchhoff's second law, Kirchhoff's loop (or mesh) rule, and Kirchhoff's second rule.

The principle of conservation of energy implies that

The directed sum of the electrical potential differences (voltage) around any closed network is zero, or:

More simply, the sum of the emfs in any closed loop is equivalent to the sum of the potential drops in that loop, or:

The algebraic sum of the products of the resistances of the conductors and the currents in them in a closed loop is equal to the total emf available in that loop.

Similarly to KCL, it can be stated as:

$$\sum_{k=1}^{n} V_k = 0$$

Here, n is the total number of voltages measured. The voltages may also be complex:

$$\sum_{k=1}^{n} \tilde{V}_k = 0$$

This law is based on the conservation of energy whereby voltage is defined as the energy per unit charge. The total amount of energy gained per unit charge must be equal to the amount of energy lost per unit charge, as energy and charge are both conserved.

Generalization

In the low-frequency limit, the voltage drop around any loop is zero. This includes imaginary loops arranged arbitrarily in space – not limited to the loops delineated by the circuit elements and conductors. In the low-frequency limit, this is a corollary of Faraday's law of induction (which is one of the Maxwell equations).

This has practical application in situations involving "static electricity".

Limitations

KCL and KVL both depend on the lumped element model being applicable to the circuit in question. When the model is not applicable, the laws do not apply.

KCL, in its usual form, is dependent on the assumption that current flows only in conductors, and that whenever current flows into one end of a conductor it immediately flows out the other end. This is not a safe assumption for high-frequency AC circuits, where the lumped element model is no longer applicable. It is often possible to improve the applicability of KCL by considering "parasitic capacitances" distributed along the conductors. Significant violations of KCL can occur even at 60 Hz, which is not a very high frequency.

In other words, KCL is valid only if the total electric charge, Q, remains constant in the region being considered. In practical cases this is always so when KCL is applied at a geometric point. When investigating a finite region, however, it is possible that the charge density within the region may change. Since charge is conserved, this can only come about by a flow of charge across the region boundary. This flow represents a net current, and KCL is violated.

KVL is based on the assumption that there is no fluctuating magnetic field linking the closed loop. This is not a safe assumption for high-frequency (short-wavelength) AC circuits. In the presence of a changing magnetic field the electric field is not a conservative vector field. Therefore, the electric field cannot be the gradient of any potential. That is to say, the line integral of the electric field around the loop is not zero, directly contradicting KVL.

It is often possible to improve the applicability of KVL by considering "parasitic inductances" (including mutual inductances) distributed along the conductors. These are treated as imaginary circuit elements that produce a voltage drop equal to the rate-of-change of the flux.

Example

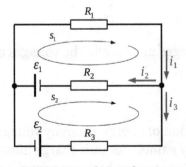

Assume an electric network consisting of two voltage sources and three resistors.

According to the first law we have

$$i_1 - i_2 - i_3 = 0$$

The second law applied to the closed circuit s_1 gives

$$-R_2 i_2 + \mathcal{E}_1 - R_1 i_1 = 0$$

The second law applied to the closed circuit s_2 gives

$$-R_3 i_3 - \mathcal{E}_2 - \mathcal{E}_1 + R_2 i_2 = 0$$

Thus we get a linear system of equations in i_1, i_2, i_3:

$$\begin{cases} i_1 - i_2 - i_3 & = 0 \\ -R_2 i_2 + \mathcal{E}_1 - R_1 i_1 & = 0 \\ -R_3 i_3 - \mathcal{E}_2 - \mathcal{E}_1 + R_2 i_2 & = 0 \end{cases}$$

Which is equivalent to

$$\begin{cases} i_1 + -i_2 + -i_3 & = 0 \\ R_1 i_1 + R_2 i_2 + 0 i_3 & = \mathcal{E}_1 \\ 0 i_1 + R_2 i_2 - R_3 i_3 & = \mathcal{E}_1 + \mathcal{E}_2 \end{cases}$$

Assuming

$$R_1 = 100, R_2 = 200, R_3 = 300 \text{ (ohms)}; \mathcal{E}_1 = 3, \mathcal{E}_2 = 4 \text{ (volts)}$$

the solution is

$$\begin{cases} i_1 = \dfrac{1}{1100} \\ i_2 = \dfrac{4}{275} \\ i_3 = -\dfrac{3}{220} \end{cases}$$

i_3 has a negative sign, which means that the direction of i_3 is opposite to the assumed direction (the direction defined in the picture).

Faraday's Law of Induction

Faraday's law of induction is a basic law of electromagnetism predicting how a magnetic field will interact with an electric circuit to produce an electromotive force (EMF)—a phenomenon called electromagnetic induction. It is the fundamental operating principle of transformers, inductors, and many types of electrical motors, generators and solenoids.

The Maxwell–Faraday equation is a generalization of Faraday's law, and is listed as one of Maxwell's equations.

History

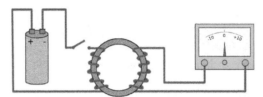

A diagram of Faraday's iron ring apparatus. Change in the magnetic flux of the left coil induces a current in the right coil.

Faraday's disk

Electromagnetic induction was discovered independently by Michael Faraday in 1831 and Joseph Henry in 1832. Faraday was the first to publish the results of his experiments. In Faraday's first experimental demonstration of electromagnetic induction (August 29, 1831), he wrapped two wires around opposite sides of an iron ring (torus) (an arrangement similar to a modern toroidal transformer). Based on his assessment of recently discovered properties of electromagnets, he expected that when current started to flow in one wire, a sort of wave would travel through the ring and cause some electrical effect on the opposite side. He plugged one wire into a galvanometer, and watched it as he connected the other wire to a battery. Indeed, he saw a transient current (which he called a "wave of electricity") when he connected the wire to the battery, and another when he disconnected it. This induction was due to the change in magnetic flux that occurred when the battery was connected and disconnected. Within two months, Faraday had found several other manifestations of electromagnetic induction. For example, he saw transient currents when he quickly slid a bar magnet in and out of a coil of wires, and he generated a steady (DC) current by rotating a copper disk near the bar magnet with a sliding electrical lead ("Faraday's disk").

Michael Faraday explained electromagnetic induction using a concept he called lines of force. However, scientists at the time widely rejected his theoretical ideas, mainly because they were not formulated mathematically. An exception was James Clerk Maxwell, who used Faraday's ideas as the basis of his quantitative electromagnetic theory. In Maxwell's papers, the time-varying aspect of electromagnetic induction is expressed as a differential equation which Oliver Heaviside referred to as Faraday's law even though it is different from the original version of Faraday's law, and does not describe motional EMF. Heaviside's version is the form recognized today in the group of equations known as Maxwell's equations.

Lenz's law, formulated by Heinrich Lenz in 1834, describes "flux through the circuit", and gives the direction of the induced EMF and current resulting from electromagnetic induction (elaborated upon in the examples below).

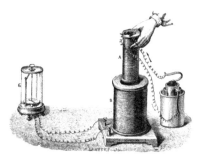

Faraday's experiment showing induction between coils of wire: The liquid battery *(right)* provides a current which flows through the small coil *(A)*, creating a magnetic field. When the coils are stationary, no current is induced. But when the small coil is moved in or out of the large coil *(B)*, the magnetic flux through the large coil changes, inducing a current which is detected by the galvanometer *(G)*.

Faraday's Law

Qualitative Statement

The most widespread version of Faraday's law states:

The induced electromotive force in any closed circuit is equal to the negative of the time rate of change of the magnetic flux enclosed by the circuit.

This version of Faraday's law strictly holds only when the closed circuit is a loop of infinitely thin wire, and is invalid in other circumstances as discussed below. A different version, the Maxwell–Faraday equation (discussed below), is valid in all circumstances.

Quantitative

The definition of surface integral relies on splitting the surface Σ into small surface elements. Each element is associated with a vector $d\mathbf{A}$ of magnitude equal to the area of the element and with direction normal to the element and pointing "outward" (with respect to the orientation of the surface).

Faraday's law of induction makes use of the magnetic flux Φ_B through a hypothetical surface Σ whose boundary is a wire loop. Since the wire loop may be moving, we write $\Sigma(t)$ for the surface. The magnetic flux is defined by a surface integral:

$$\Phi_B = \iint_{\Sigma(t)} B(r,t) \cdot dA,$$

where dA is an element of surface area of the moving surface $\Sigma(t)$, B is the magnetic field (also called "magnetic flux density"), and B·dA is a vector dot product (the infinitesimal amount of magnetic flux through the infinitesimal area element dA). In more visual terms, the magnetic flux through the wire loop is proportional to the number of magnetic flux lines that pass through the loop.

When the flux changes—because B changes, or because the wire loop is moved or deformed, or both—Faraday's law of induction says that the wire loop acquires an EMF, $\mathcal{E}$, defined as the energy available from a unit charge that has travelled once around the wire loop. Equivalently, it is the voltage that would be measured by cutting the wire to create an open circuit, and attaching a voltmeter to the leads.

Faraday's law states that the EMF is also given by the rate of change of the magnetic flux:

$$\mathcal{E} = -\frac{d\Phi_B}{dt},$$

where $\mathcal{E}$, is the electromotive force (EMF) and Φ_B is the magnetic flux. The direction of the electromotive force is given by Lenz's law.

For a tightly wound coil of wire, composed of N identical turns, each with the same Φ_B, Faraday's law of induction states that

$$\mathcal{E} = -N\frac{d\Phi_B}{dt}$$

where N is the number of turns of wire and Φ_B is the magnetic flux through a *single* loop.

Maxwell–Faraday Equation

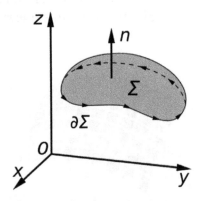

An illustration of Kelvin–Stokes theorem with surface Σ its boundary $\partial\Sigma$ and orientation n set by the right-hand rule.

The Maxwell–Faraday equation is a generalisation of Faraday's law that states that a time-varying magnetic field will always accompany a spatially-varying, non-conservative electric field, and vice versa. The Maxwell–Faraday equation is

$$\nabla \times \mathbf{E} = -\frac{\partial \mathbf{B}}{\partial t}$$

(in SI units) where $\nabla \times$ is the curl operator and again E(r, t) is the electric field and B(r, t) is the magnetic field. These fields can generally be functions of position r and time t.

The Maxwell–Faraday equation is one of the four Maxwell's equations, and therefore plays a fundamental role in the theory of classical electromagnetism. It can also be written in an integral form by the Kelvin–Stokes theorem:

$$\oint_{\partial \Sigma} \mathbf{E} \cdot d\mathbf{l} = -\int_{\Sigma} \frac{\partial \mathbf{B}}{\partial t} \cdot d\mathbf{A}$$

where, as indicated in the figure:

 Σ is a surface bounded by the closed contour $\partial \Sigma$,

 E is the electric field, B is the magnetic field.

 dℓ is an infinitesimal vector element of the contour $\partial \Sigma$,

 dA is an infinitesimal vector element of surface Σ. If its direction is orthogonal to that surface patch, the magnitude is the area of an infinitesimal patch of surface.

Both dℓ and dA have a sign ambiguity; to get the correct sign, the right-hand rule is used, as explained in the article Kelvin–Stokes theorem. For a planar surface Σ, a positive path element $d\ell$ of curve $\partial \Sigma$ is defined by the right-hand rule as one that points with the fingers of the right hand when the thumb points in the direction of the normal n to the surface Σ.

The integral around $\partial \Sigma$ is called a *path integral* or *line integral.*

Notice that a nonzero path integral for E is different from the behavior of the electric field generated by charges. A charge-generated E-field can be expressed as the gradient of a scalar field that is a solution to Poisson's equation, and has a zero path integral.

The integral equation is true for *any* path $\partial \Sigma$ through space, and any surface Σ for which that path is a boundary.

If the path Σ is not changing in time, the equation can be rewritten:

$$\oint_{\partial \Sigma} \mathbf{E} \cdot d\mathbf{l} = -\frac{d}{dt} \int_{\Sigma} \mathbf{B} \cdot d\mathbf{A}.$$

The surface integral at the right-hand side is the explicit expression for the magnetic flux Φ_B through Σ.

Proof of Faraday's Law

The four Maxwell's equations (including the Maxwell–Faraday equation), along with the Lorentz force law, are a sufficient foundation to derive *everything* in classical electromagnetism. Therefore, it is possible to "prove" Faraday's law starting with these equations. Click "show" in the box below for an outline of this proof. (In an alternative approach, not shown here but equally valid, Faraday's law could be taken as the starting point and used to "prove" the Maxwell–Faraday equation and/or other laws.)

"Counterexamples" to Faraday's Law

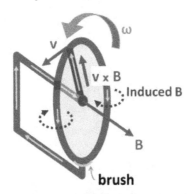

Faraday's disc electric generator. The disc rotates with angular rate ω, sweeping the conducting radius circularly in the static magnetic field B. The magnetic Lorentz force v × B drives the current along the conducting radius to the conducting rim, and from there the circuit completes through the lower brush and the axle supporting the disc. This device generates an EMF and a current, although the shape of the "circuit" is constant and thus the flux through the circuit does not change with time.

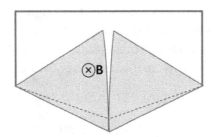

A counterexample to Faraday's law when over-broadly interpreted. A wire (solid red lines) connects to two touching metal plates (silver) to form a circuit. The whole system sits in a uniform magnetic field, normal to the page. If the word "circuit" is interpreted as "primary path of current flow" (marked in red), then the magnetic flux through the "circuit" changes dramatically as the plates are rotated, yet the EMF is almost zero, which contradicts Faraday's law. After *Feynman Lectures on Physics* Vol. II page 17-3.

Although Faraday's law is always true for loops of thin wire, it can give the wrong result if naively extrapolated to other contexts. One example is the homopolar generator (above left): A spinning circular metal disc in a homogeneous magnetic field generates a DC (constant in time) EMF. In Faraday's law, EMF is the time-derivative of flux, so a DC EMF is only possible if the magnetic flux is getting uniformly larger and larger perpetually. But in the generator, the magnetic field is constant and the disc stays in the same position, so no magnetic fluxes are growing larger and larger. So this example cannot be analyzed directly with Faraday's law.

Another example, due to Feynman, has a dramatic change in flux through a circuit, even though the EMF is arbitrarily small.

In both these examples, the changes in the current path are different from the motion of the material making up the circuit. The electrons in a material tend to follow the motion of the atoms that make up the material, due to scattering in the bulk and work function confinement at the edges.

Therefore, motional EMF is generated when a material's atoms are moving through a magnetic field, dragging the electrons with them, thus subjecting the electrons to the Lorentz force. In the homopolar generator, the material's atoms are moving, even though the overall geometry of the circuit is staying the same. In the second example, the material's atoms are almost stationary, even though the overall geometry of the circuit is changing dramatically. On the other hand, Faraday's law always holds for thin wires, because there the geometry of the circuit always changes in a direct relationship to the motion of the material's atoms.

Although Faraday's law does not apply to all situations, the Maxwell–Faraday equation and Lorentz force law are always correct and can always be used directly.

Both of the above examples can be correctly worked by choosing the appropriate path of integration for Faraday's law. Outside of context of thin wires, the path must never be chosen to go through the conductor in the shortest direct path. This is explained in detail in "The Electromagnetodynamics of Fluid" by W. F. Hughes and F. J. Young, John Wiley Inc. (1965).

Faraday's Law and Relativity

Two Phenomena

Some physicists have remarked that Faraday's law is a single equation describing two different phenomena: the *motional EMF* generated by a magnetic force on a moving wire and the *transformer EMF* generated by an electric force due to a changing magnetic field (due to the Maxwell–Faraday equation).

James Clerk Maxwell drew attention to this fact in his 1861 paper *On Physical Lines of Force*. In the latter half of Part II of that paper, Maxwell gives a separate physical explanation for each of the two phenomena.

A reference to these two aspects of electromagnetic induction is made in some modern textbooks. As Richard Feynman states:

So the "flux rule" that the emf in a circuit is equal to the rate of change of the magnetic flux through the circuit applies whether the flux changes because the field changes or because the circuit moves (or both) ...

Yet in our explanation of the rule we have used two completely distinct laws for the two cases- $v \times B$ for "circuit moves" and $\nabla \times E = -\partial_t B$ for "field changes".

We know of no other place in physics where such a simple and accurate general principle requires for its real understanding an analysis in terms of *two different phenomena*.

— *Richard P. Feynman, The Feynman Lectures on Physics*

Einstein's View

Reflection on this apparent dichotomy was one of the principal paths that led Einstein to develop special relativity:

It is known that Maxwell's electrodynamics—as usually understood at the present time—when applied to moving bodies, leads to asymmetries which do not appear to be inherent in the phenomena. Take, for example, the reciprocal electrodynamic action of a magnet and a conductor.

The observable phenomenon here depends only on the relative motion of the conductor and the magnet, whereas the customary view draws a sharp distinction between the two cases in which either the one or the other of these bodies is in motion. For if the magnet is in motion and the conductor at rest, there arises in the neighbourhood of the magnet an electric field with a certain definite energy, producing a current at the places where parts of the conductor are situated.

But if the magnet is stationary and the conductor in motion, no electric field arises in the neighbourhood of the magnet. In the conductor, however, we find an electromotive force, to which in itself there is no corresponding energy, but which gives rise—assuming equality of relative motion in the two cases discussed—to electric currents of the same path and intensity as those produced by the electric forces in the former case.

Examples of this sort, together with unsuccessful attempts to discover any motion of the earth relative to the "light medium," suggest that the phenomena of electrodynamics as well as of mechanics possess no properties corresponding to the idea of absolute rest.

—Albert Einstein, On the Electrodynamics of Moving Bodies

Coulomb's Law

Coulomb's law or Coulomb's inverse-square law, is a law of physics that describes force interacting between static electrically charged particles. In its scalar form the law is:

$$F = k_e \frac{q_1 q_2}{r^2},$$

where k_e is Coulomb's constant ($k_e = 8.99 \times 10^9$ N m^2 C^{-2}), q_1 and q_2 are the signed magnitudes of the charges, and the scalar r is the distance between the charges. The force of interaction between the charges is attractive if the charges have opposite signs (i.e. F is negative) and repulsive if like-signed (i.e. F is positive).

The law was first published in 1784 by French physicist Charles Augustin de Coulomb and was essential to the development of the theory of electromagnetism. It is analogous to Isaac Newton's inverse-square law of universal gravitation. Coulomb's law can be used to derive Gauss's law, and vice versa. The law has been tested extensively, and all observations have upheld the law's principle.

History

Ancient cultures around the Mediterranean knew that certain objects, such as rods of amber, could be rubbed with cat's fur to attract light objects like feathers. Thales of Miletus made a series of observations on static electricity around 600 BC, from which he believed that friction rendered amber magnetic, in contrast to minerals such as magnetite, which needed no rubbing. Thales was incorrect in believing the attraction was due to a magnetic effect, but later science would prove a link between magnetism and electricity. Electricity would remain little more than an intellectual curiosity for millennia until 1600, when the English scientist William Gilbert made a careful study of electricity and magnetism, distinguishing the lodestone effect from static electricity produced by rubbing amber. He coined the New Latin word *electricus* to refer to the property of attracting small objects after being rubbed. This association gave rise to the English words "electric" and "electricity", which made their first appearance in print in Thomas Browne's *Pseudodoxia Epidemica* of 1646.

Charles-Augustin de Coulomb

Early investigators of the 18th century who suspected that the electrical force diminished with distance as the force of gravity did (i.e., as the inverse square of the distance) included Daniel Bernoulli and Alessandro Volta, both of whom measured the force between plates of a capacitor, and Franz Aepinus who supposed the inverse-square law in 1758.

Based on experiments with electrically charged spheres, Joseph Priestley of England was among the first to propose that electrical force followed an inverse-square law, similar to Newton's law of universal gravitation. However, he did not generalize or elaborate on this. In 1767, he conjectured that the force between charges varied as the inverse square of the distance.

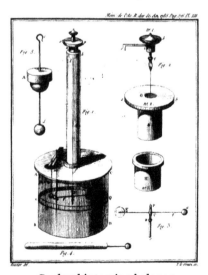

Coulomb's torsion balance

In 1769, Scottish physicist John Robison announced that, according to his measurements, the force of repulsion between two spheres with charges of the same sign varied as $x^{-2.06}$.

In the early 1770s, the dependence of the force between charged bodies upon both distance and charge had already been discovered, but not published, by Henry Cavendish of England.

Finally, in 1785, the French physicist Charles-Augustin de Coulomb published his first three reports of electricity and magnetism where he stated his law. This publication was essential to the development of the theory of electromagnetism. He used a torsion balance to study the repulsion and attraction forces of charged particles, and determined that the magnitude of the electric force between two point charges is directly proportional to the product of the charges and inversely proportional to the square of the distance between them.

The torsion balance consists of a bar suspended from its middle by a thin fiber. The fiber acts as a very weak torsion spring. In Coulomb's experiment, the torsion balance was an insulating rod with a metal-coated ball attached to one end, suspended by a silk thread. The ball was charged with a known charge of static electricity, and a second charged ball of the same polarity was brought near it. The two charged balls repelled one another, twisting the fiber through a certain angle, which could be read from a scale on the instrument. By knowing how much force it took to twist the fiber through a given angle, Coulomb was able to calculate the force between the balls and derive his inverse-square proportionality law.

The Law

Coulomb's law states that:

The magnitude of the electrostatic force of interaction between two point charges is directly proportional to the scalar multiplication of the magnitudes of charges and inversely proportional to the square of the distance between them.

The force is along the straight line joining them. If the two charges have the same sign, the electrostatic force between them is repulsive; if they have different signs, the force between them is attractive.

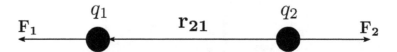

Coulomb's law can also be stated as a simple mathematical expression. The scalar and vector forms of the mathematical equation are

$$|\mathbf{F}| = k_e \frac{|q_1 q_2|}{r^2} \text{ and } \mathbf{F}_1 = k_e \frac{q_1 q_2}{|\mathbf{r}_{12}|^2} \hat{\mathbf{r}}_{21}, \text{ respectively,}$$

where k_e is Coulomb's constant ($k_e = 8.9875517873681764 \times 10^9$ N m² C⁻²), q_1 and q_2 are the signed magnitudes of the charges, the scalar r is the distance between the charges, the vector $\mathbf{r}_{21} = \mathbf{r}_1 - \mathbf{r}_2$ is the vectorial distance between the charges, and $\hat{\mathbf{r}}_{21} = \mathbf{r}_{21}/|\mathbf{r}_{21}|$ (a unit vector pointing from q_2 to q_1). The vector form of the equation calculates the force $\mathbf{F}_1$ applied on q_1 by q_2. If $\mathbf{r}_{12}$ is used instead, then the effect on q_2 can be found. It can be also calculated using Newton's third law: $\mathbf{F}_2 = -\mathbf{F}_1$.

Units

Electromagnetic theory is usually expressed using the standard SI units. Force is measured in newtons, charge in coulombs, and distance in metres. Coulomb's constant is given by $k_e = \frac{1}{4\pi\varepsilon_0}$. The constant ε_0 is the permittivity of free space in $C^2\ m^{-2}\ N^{-1}$. And ε is the relative permittivity of the material in which the charges are immersed, and is dimensionless.

The SI derived units for the electric field are volts per meter, newtons per coulomb, or tesla meters per second.

Coulomb's law and Coulomb's constant can also be interpreted in various terms:

- Atomic units. In atomic units the force is expressed in hartrees per Bohr radius, the charge in terms of the elementary charge, and the distances in terms of the *Bohr radius*.

- Electrostatic units or Gaussian units. In electrostatic units and Gaussian units, the unit charge (*esu* or statcoulomb) is defined in such a way that the Coulomb constant k disappears because it has the value of one and becomes dimensionless.

Electric Field

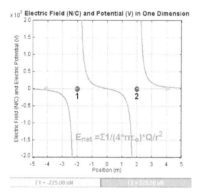

If the two charges have the same sign, the electrostatic force between them is repulsive; if they have different sign, the force between them is attractive.

An electric field is a vector field that associates to each point in space the Coulomb force experienced by a test charge. In the simplest case, the field is considered to be generated solely by a single source point charge. The strength and direction of the Coulomb force F on a test charge q_t depends on the electric field E that it finds itself in, such that $F = q_t E$. If the field is generated by a positive source point charge q, the direction of the electric field points along lines directed radially outwards from it, i.e. in the direction that a positive point test charge q_t would move if placed in the field. For a negative point source charge, the direction is radially inwards.

The magnitude of the electric field E can be derived from Coulomb's law. By choosing one of the point charges to be the source, and the other to be the test charge, it follows from Coulomb's law that the magnitude of the electric field E created by a single source point charge q at a certain distance from it r in vacuum is given by:

$$|\mathbf{E}| = \frac{1}{4\pi\varepsilon_0}\frac{|q|}{r^2}$$

Coulomb's Constant

Coulomb's constant is a proportionality factor that appears in Coulomb's law as well as in other electric-related formulas. Denoted k_e, it is also called the electric force constant or electrostatic constant, hence the subscript e.

The exact value of Coulomb's constant is:

$$k_e = \frac{1}{4\pi\varepsilon_0} = \frac{c_0^2 \mu_0}{4\pi} = c_0^2 \times 10^{-7} \ \text{H}\cdot\text{m}^{-1}$$
$$= 8.9875517873681764 \times 10^9 \ \text{N}\cdot\text{m}^2\cdot\text{C}^{-2}$$

Conditions for Validity

There are three conditions to be fulfilled for the validity of Coulomb's law:

- The charges must have a spherically symmetric distribution (e.g. be point charges, or a charged metal sphere.

- The charges must not overlap (e.g. be distinct point charges)

- The charges must be stationary with respect to each other.

Scalar Form

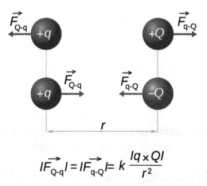

$$|\vec{F}_{Q\text{-}q}| = |\vec{F}_{q\text{-}Q}| = k\,\frac{|q \times Q|}{r^2}$$

The absolute value of the force **F** between two point charges q and Q relates to the distance between the point charges and to the simple product of their charges. The diagram shows that like charges repel each other, and opposite charges attract each other.

When it is only of interest to know the magnitude of the electrostatic force (and not its direction), it may be easiest to consider a scalar version of the law. The scalar form of Coulomb's Law relates the magnitude and sign of the electrostatic force F acting simultaneously on two point charges q_1 and q_2 as follows:

$$|\mathbf{F}| = k_e\,\frac{|q_1 q_2|}{r^2}$$

where r is the separation distance and k_e is Coulomb's constant. If the product $q_1 q_2$ is positive, the force between the two charges is repulsive; if the product is negative, the force between them is attractive.

Vector Form

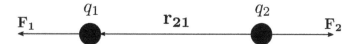

In the image, the vector F_1 is the force experienced by q_1, and the vector F_2 is the force experienced by q_2. When $q_1 q_2 > 0$ *the forces are repulsive (as in the image) and when* $q_1 q_2 < 0$ *the forces are attractive (opposite to the image). The magnitude of the forces will always be equal.*

Coulomb's law states that the electrostatic force F_1 experienced by a charge, q_1 at position r_1, in the vicinity of another charge, q_2 at position r_2, in a vacuum is equal to:

$$F_1 = \frac{q_1 q_2}{4\pi\varepsilon_0} \frac{(r_1 - r_2)}{|r_1 - r_2|^3} = \frac{q_1 q_2}{4\pi\varepsilon_0} \frac{\hat{r}_{21}}{|r_{21}|^2}$$

where $r_{21} = r_1 - r_2$, the unit vector $\hat{r}_{21} = r_{21}/|r_{21}|$, and ε_0 is the electric constant.

The vector form of Coulomb's law is simply the scalar definition of the law with the direction given by the unit vector, $\hat{r}_{21}$, parallel with the line *from* charge q_2 *to* charge q_1. If both charges have the same sign (like charges) then the product $q_1 q_2$ is positive and the direction of the force on q_1 is given by $\hat{r}_{21}$; the charges repel each other. If the charges have opposite signs then the product $q_1 q_2$ is negative and the direction of the force on q_1 is given by $-\hat{r}_{21} = \hat{r}_{12}$; the charges attract each other.

The electrostatic force F_2 experienced by q_2, according to Newton's third law, is $F_2 = -F_1$.

System of Discrete Charges

The law of superposition allows Coulomb's law to be extended to include any number of point charges. The force acting on a point charge due to a system of point charges is simply the vector addition of the individual forces acting alone on that point charge due to each one of the charges. The resulting force vector is parallel to the electric field vector at that point, with that point charge removed.

The force F on a small charge q at position r, due to a system of N discrete charges in vacuum is:

$$F(r) = \frac{q}{4\pi\varepsilon_0} \sum_{i=1}^{N} q_i \frac{r - r_i}{|r - r_i|^3} = \frac{q}{4\pi\varepsilon_0} \sum_{i=1}^{N} q_i \frac{\hat{R}_i}{|R_i|^2},$$

where q_i and r_i are the magnitude and position respectively of the ith charge, $\hat{R}_i$ is a unit vector in the direction of $R_i = r - r_i$ (a vector pointing from charges q_i to q).

Continuous Charge Distribution

In this case, the principle of linear superposition is also used. For a continuous charge distribution, an integral over the region containing the charge is equivalent to an infinite summation, treating each infinitesimal element of space as a point charge dq. The distribution of charge is usually linear, surface or volumetric.

For a linear charge distribution (a good approximation for charge in a wire) where $\lambda(\mathbf{r}')$ gives the charge per unit length at position r', and dl' is an infinitesimal element of length,

$$dq = \lambda(\mathbf{r}')dl'.$$

For a surface charge distribution (a good approximation for charge on a plate in a parallel plate capacitor) where $\sigma(\mathbf{r}')$ gives the charge per unit area at position r', and dA' is an infinitesimal element of area,

$$dq = \sigma(\mathbf{r}')dA'.$$

For a volume charge distribution (such as charge within a bulk metal) where $\rho(\mathbf{r}')$ gives the charge per unit volume at position r', and dV' is an infinitesimal element of volume,

$$dq = \rho(\mathbf{r}')dV'.$$

The force on a small test charge q' at position r in vacuum is given by the integral over the distribution of charge:

$$\mathbf{F} = \frac{q'}{4\pi\varepsilon_0}\int dq \frac{\mathbf{r}-\mathbf{r}'}{|\mathbf{r}-\mathbf{r}'|^3}.$$

Simple Experiment to Verify Coulomb's Law

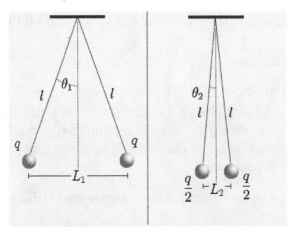

Experiment to verify Coulomb's law.

It is possible to verify Coulomb's law with a simple experiment. Consider two small spheres of mass m and same-sign charge q, hanging from two ropes of negligible mass of length l. The forces acting on each sphere are three: the weight mg, the rope tension T and the electric force F.

In the Equilibrium State:

$$T \sin\theta_1 = F_1 \tag{1}$$

and:

$$T \cos\theta_1 = mg \tag{2}$$

Dividing (1) by (2):

$$\frac{\sin\theta_1}{\cos\theta_1} = \frac{F_1}{mg} \Rightarrow F_1 = mg\tan\theta_1 \qquad (3)$$

Let L_1 be the distance between the charged spheres; the repulsion force between them F_1, assuming Coulomb's law is correct, is equal to

$$F_1 = \frac{q^2}{4\pi\epsilon_0 L_1^2} \qquad \textbf{\textit{(Coulomb's law)}}$$

so:

$$\frac{q^2}{4\pi\epsilon_0 L_1^2} = mg\tan\theta_1 \qquad (4)$$

If we now discharge one of the spheres, and we put it in contact with the charged sphere, each one of them acquires a charge $q/2$. In the equilibrium state, the distance between the charges will be $L_2 < L_1$ and the repulsion force between them will be:

$$F_2 = \frac{(\frac{q}{2})^2}{4\pi\epsilon_0 L_2^2} = \frac{\frac{q^2}{4}}{4\pi\epsilon_0 L_2^2} \qquad (5)$$

We now that $F_2 = mg\tan\theta_2$. And:

$$\frac{\frac{q^2}{4}}{4\pi\epsilon_0 L_2^2} = mg\tan\theta_2$$

Dividing (4) by (5), we get:

$$\frac{\left(\dfrac{q^2}{4\pi\epsilon_0 L_1^2}\right)}{\left(\dfrac{\frac{q^2}{4}}{4\pi\epsilon_0 L_2^2}\right)} = \frac{mg\tan\theta_1}{mg\tan\theta_2} \Rightarrow 4\left(\frac{L_2}{L_1}\right)^2 = \frac{\tan\theta_1}{\tan\theta_2} \qquad (6)$$

Measuring the angles θ_1 and θ_2 and the distance between the charges L_1 and L_2 is sufficient to verify that the equality is true taking into account the experimental error. In practice, angles can be difficult to measure, so if the length of the ropes is sufficiently great, the angles will be small enough to make the following approximation:

$$\tan\theta \approx \sin\theta = \frac{\frac{L}{2}}{l} = \frac{L}{2l} \Rightarrow \frac{\tan\theta_1}{\tan\theta_2} \approx \frac{\frac{L_1}{2l}}{\frac{L_2}{2l}} \qquad (7)$$

Using this approximation, the relationship (6) becomes the much simpler expression:

$$\frac{\frac{L_1}{2l}}{\frac{L_2}{2l}} \approx 4\left(\frac{L_2}{L_1}\right)^2 \Rightarrow \frac{L_1}{L_2} \approx 4\left(\frac{L_2}{L_1}\right)^2 \Rightarrow \frac{L_1}{L_2} \approx \sqrt[3]{4} \qquad (8)$$

In this way, the verification is limited to measuring the distance between the charges and check that the division approximates the theoretical value.

Electrostatic Approximation

In either formulation, Coulomb's law is fully accurate only when the objects are stationary, and remains approximately correct only for slow movement. These conditions are collectively known as the electrostatic approximation. When movement takes place, magnetic fields that alter the force on the two objects are produced. The magnetic interaction between moving charges may be thought of as a manifestation of the force from the electrostatic field but with Einstein's theory of relativity taken into consideration.

Atomic Forces

Coulomb's law holds even within atoms, correctly describing the force between the positively charged atomic nucleus and each of the negatively charged electrons. This simple law also correctly accounts for the forces that bind atoms together to form molecules and for the forces that bind atoms and molecules together to form solids and liquids. Generally, as the distance between ions increases, the energy of attraction approaches zero and ionic bonding is less favorable. As the magnitude of opposing charges increases, energy increases and ionic bonding is more favorable.

Gauss's Law

In physics, Gauss's law, also known as Gauss's flux theorem, is a law relating the distribution of electric charge to the resulting electric field.

The law was first formulated by Joseph-Louis Lagrange in 1773, followed by Carl Friedrich Gauss in 1813, both in the context of the attraction of ellipsoids. It is one of Maxwell's four equations, which form the basis of classical electrodynamics. Gauss's law can be used to derive Coulomb's law, and vice versa.

Qualitative Description

In words, Gauss's law states that:

> *The net electric flux through any closed surface is equal to $\frac{1}{\varepsilon}$ times the net electric charge within that closed surface.*

Gauss' law has a close mathematical similarity with a number of laws in other areas of physics, such as Gauss' law for magnetism and Gauss' law for gravity. In fact, any "inverse-square law" can be formulated in a way similar to Gauss' law: For example, Gauss' law itself is essentially equivalent to the inverse-square Coulomb's law, and Gauss' law for gravity is essentially equivalent to the inverse-square Newton's law of gravity.

Gauss' law is something of an electrical analogue of Ampère's law, which deals with magnetism.

The law can be expressed mathematically using vector calculus in integral form and differential form, both are equivalent since they are related by the divergence theorem, also called Gauss' theorem. Each of these forms in turn can also be expressed two ways: In terms of a relation between the electric field E and the total electric charge, or in terms of the electric displacement field D and the *free* electric charge.

Equation Involving E Field

Gauss's law can be stated using either the electric field E or the electric displacement field D. This section shows some of the forms with E; the form with D is below, as are other forms with E.

Integral Form

Gauss's law may be expressed as:

$$\Phi_E = \frac{Q}{\varepsilon_0}$$

where Φ_E is the electric flux through a closed surface S enclosing any volume V, Q is the total charge enclosed within S, and ε_0 is the electric constant. The electric flux Φ_E is defined as a surface integral of the electric field:

$$\phi_E = \oiint_S E.dA$$

where E is the electric field, dA is a vector representing an infinitesimal element of area of the surface, and · represents the dot product of two vectors.

Since the flux is defined as an *integral* of the electric field, this expression of Gauss's law is called the *integral form*.

Applying the Integral Form

If the electric field is known everywhere, Gauss's law makes it quite easy, in principle, to find the distribution of electric charge: The charge in any given region can be deduced by integrating the electric field to find the flux.

However, much more often, it is the reverse problem that needs to be solved: The electric charge distribution is known, and the electric field needs to be computed. This is much more difficult, since if you know the total flux through a given surface, that gives almost no information about the electric field, which (for all you know) could go in and out of the surface in arbitrarily complicated patterns.

An exception is if there is some symmetry in the situation, which mandates that the electric field passes through the surface in a uniform way. Then, if the total flux is known, the field itself can be deduced at every point. Common examples of symmetries which lend themselves to Gauss's law include cylindrical symmetry, planar symmetry, and spherical symmetry.

Differential Form

By the divergence theorem, Gauss's law can alternatively be written in the *differential form*:

$$\nabla \cdot \mathbf{E} = \frac{\rho}{\varepsilon_0}$$

where $\nabla \cdot E$ is the divergence of the electric field, ε_0 is the electric constant, and ρ is the total electric charge density (charge per unit volume).

Equivalence of Integral and Differential Forms

The integral and differential forms are mathematically equivalent, by the divergence theorem. Here is the argument more specifically.

The integral form of Gauss' law is:

$$\oiint_S E.dA = \frac{Q}{\in_0}$$

for any closed surface S containing charge Q. By the divergence theorem, this equation is equivalent to:

$$\iiint_v \nabla.EdA = \frac{Q}{\in_0}$$

for any volume V containing charge Q. By the relation between charge and charge density, this equation is equivalent to:

$$\iiint_v \nabla.EdA = \iiint_v \frac{\rho}{\in_0}dv$$

for any volume V. In order for this equation to be simultaneously true for every possible volume V, it is necessary (and sufficient) for the integrands to be equal everywhere. Therefore, this equation is equivalent to:

$$\nabla.E = \frac{\rho}{\in_0}$$

Thus the integral and differential forms are equivalent

Equation Involving D Field

Free, Bound, and Total Charge

The electric charge that arises in the simplest textbook situations would be classified as "free charge"—for example, the charge which is transferred in static electricity, or the charge on a capacitor plate. In contrast, "bound charge" arises only in the context of dielectric (polarizable) materials. (All materials are polarizable to some extent.) When such materials are placed in an external electric field, the electrons remain bound to their respective atoms, but shift a microscopic distance in response to the field, so that they're more on one side of the atom than the other. All these microscopic displacements add up to give a macroscopic net charge distribution, and this constitutes the "bound charge".

Although microscopically, all charge is fundamentally the same, there are often practical reasons for wanting to treat bound charge differently from free charge. The result is that the more "fundamental" Gauss's law, in terms of E (above), is sometimes put into the equivalent form below, which is in terms of D and the free charge only.

Integral Form

This formulation of Gauss's law states the total charge form:

$$\Phi_D = Q_{free}$$

where Φ_D is the D-field flux through a surface S which encloses a volume V, and Q_{free} is the free charge contained in V. The flux Φ_D is defined analogously to the flux Φ_E of the electric field E through S:

$$\phi_D = \oiint_S D.dA$$

Differential Form

The differential form of Gauss's law, involving free charge only, states:

$$\nabla \cdot \mathbf{D} = \rho_{free}$$

where $\nabla \cdot D$ is the divergence of the electric displacement field, and ρ_{free} is the free electric charge density.

Equivalence of Total and Free Charge Statements

Proof that the formulations of Gauss' law in terms of free charge are equivalent to the formulations involving total charge.

In this proof, we will show that the equation

$$\nabla \cdot \mathbf{E} = \frac{\rho}{\varepsilon_0}$$

is equivalent to the equation

$$\nabla \cdot \mathbf{D} = \rho_{free}$$

Note that we are only dealing with the differential forms, not the integral forms, but that is sufficient since the differential and integral forms are equivalent in each case, by the divergence theorem.

We introduce the polarization density P, which has the following relation to E and D:

$$\mathbf{D} = \varepsilon_0 \mathbf{E} + \mathbf{P}$$

and the following relation to the bound charge:

$$\rho_{bound} = -\nabla \cdot \mathbf{P}$$

Now, consider the three equations:

$$\rho_{bound} = \nabla \cdot (-\mathbf{P})$$

$$\rho_{free} = \nabla \cdot \mathbf{D}$$

$$\rho = \nabla \cdot (\varepsilon_0 \mathbf{E})$$

The key insight is that the sum of the first two equations is the third equation. This completes the proof: The first equation is true by definition, and therefore the second equation is true if and only if the third equation is true. So the second and third equations are equivalent, which is what we wanted to prove.

Equation for Linear Materials

In homogeneous, isotropic, nondispersive, linear materials, there is a simple relationship between E and D:

$$\mathbf{D} = \varepsilon \mathbf{E}$$

where ε is the permittivity of the material. For the case of vacuum (aka free space), $\varepsilon = \varepsilon_0$. Under these circumstances, Gauss's law modifies to

$$\Phi_E = \frac{Q_{free}}{\varepsilon}$$

for the integral form, and

$$\nabla \cdot \mathbf{E} = \frac{\rho_{free}}{\varepsilon}$$

for the differential form.

Interpretations

In Terms of Fields of Force

Gauss' theorem can be interpreted in terms of the lines of force of the field as follows:

The flow field through the surface is The number of field lines penetrating the surface. This takes into account the direction of - field lines penetrating the surface considered with a minus sign in the opposite direction. Force lines begin or end only charges (start on the positive end to the negative), or may even go to infinity. The number of lines of force emanating from the charge (starting it) anyway , the magnitude of this charge (the charge is defined in the model). (For all the negative charges of the same, only the charge is equal to minus the number of its member (it ends) lines. On the basis of these two provisions of the Gauss theorem is evident in the statement: the number of lines emanating from a closed surface is equal to the total number of charges inside it - that is, the number of lines that appear within it . Of course, meant keeping signs, in particular, the line, which began within the surface on the positive charge can end on a negative charge and within it (if there is), then it does not give a contribution to the flux through this surface, as, or even before it not reach, or be released, and then enters back (or, in general, the surface intersects an even number of times equal to the forward and the opposite direction) that gives zero contribution to the flow in the summation with the correct sign. The same can be said about the lines begin and end outside the given surface - for the same reason, they also give a zero contribution to flow through it.

Relation to Coulomb's Law

Deriving Gauss's law from Coulomb's Law

Gauss's law can be derived from Coulomb's law.

Coulomb's law states that the electric field due to a stationary point charge is:

$$E(r) = \frac{q}{4\pi\varepsilon_0} \frac{e_r}{r^2}$$

where

e_r is the radial unit vector,

r is the radius, $|r|$,

ε_o is the electric constant,

q is the charge of the particle, which is assumed to be located at the origin.

Using the expression from Coulomb's law, we get the total field at r by using an integral to sum the field at r due to the infinitesimal charge at each other point s in space, to give

$$E(r) = \frac{1}{4\pi\varepsilon_0} \int \frac{\rho(s)(r-s)}{|r-s|^3} d^3s$$

where ρ is the charge density. If we take the divergence of both sides of this equation with respect to r, and use the known theorem

$$\nabla \cdot \left(\frac{\mathbf{r}}{|\mathbf{r}|^3} \right) = 4\pi\delta(\mathbf{r})$$

where $\delta(r)$ is the Dirac delta function, the result is

$$\nabla \cdot E(r) = \frac{1}{\varepsilon_0} \int \rho(s)\delta(r-s)d^3s$$

Using the "sifting property" of the Dirac delta function, we arrive at

$$\nabla \cdot E(r) = \frac{\rho(r)}{\varepsilon_0},$$

which is the differential form of Gauss's law, as desired.

Note that since Coulomb's law only applies to *stationary* charges, there is no reason to expect Gauss's law to hold for moving charges *based on this derivation alone*. In fact, Gauss's law *does* hold for moving charges, and in this respect Gauss's law is more general than Coulomb's law.

Deriving Coulomb's law from Gauss's Law

Strictly speaking, Coulomb's law cannot be derived from Gauss's law alone, since Gauss's law does not give any information regarding the curl of E. However, Coulomb's law *can* be proven from

Gauss's law if it is assumed, in addition, that the electric field from a point charge is spherical-ly-symmetric (this assumption, like Coulomb's law itself, is exactly true if the charge is stationary, and approximately true if the charge is in motion).

Taking S in the integral form of Gauss's law to be a spherical surface of radius r, centered at the point charge Q, we have

$$\oint_S \mathbf{E} \cdot d\mathbf{A} = \frac{Q}{\varepsilon_0}$$

By the assumption of spherical symmetry, the integrand is a constant which can be taken out of the integral. The result is

$$4\pi r^2 \hat{\mathbf{r}} \cdot \mathbf{E}(\mathbf{r}) = \frac{Q}{\varepsilon_0}$$

where $\hat{r}$ is a unit vector pointing radially away from the charge. Again by spherical symmetry, E points in the radial direction, and so we get

$$\mathbf{E}(\mathbf{r}) = \frac{Q}{4\pi\varepsilon_0} \frac{\hat{r}}{r^2}$$

which is essentially equivalent to Coulomb's law. Thus the inverse-square law dependence of the electric field in Coulomb's law follows from Gauss's law.

Ohm's Law

V, I, and R, the parameters of Ohm's law.

Ohm's law states that the current through a conductor between two points is directly proportional to the voltage across the two points. Introducing the constant of proportionality, the resistance, one arrives at the usual mathematical equation that describes this relationship:

$$I = \frac{V}{R},$$

where I is the current through the conductor in units of amperes, V is the voltage measured *across* the conductor in units of volts, and R is the resistance of the conductor in units of ohms. More specifically, Ohm's law states that the R in this relation is constant, independent of the current.

The law was named after the German physicist Georg Ohm, who, in a treatise published in 1827, described measurements of applied voltage and current through simple electrical circuits containing various lengths of wire. He presented a slightly more complex equation than the one above to explain his experimental results. The above equation is the modern form of Ohm's law.

In physics, the term *Ohm's law* is also used to refer to various generalizations of the law originally formulated by Ohm. The simplest example of this is:

$$\mathbf{J} = \sigma\mathbf{E},$$

where $\mathbf{J}$ is the current density at a given location in a resistive material, $\mathbf{E}$ is the electric field at that location, and σ (Sigma) is a material-dependent parameter called the conductivity. This reformulation of Ohm's law is due to Gustav Kirchhoff.

History

Georg Ohm

In January 1781, before Georg Ohm's work, Henry Cavendish experimented with Leyden jars and glass tubes of varying diameter and length filled with salt solution. He measured the current by noting how strong a shock he felt as he completed the circuit with his body. Cavendish wrote that the "velocity" (current) varied directly as the "degree of electrification" (voltage). He did not communicate his results to other scientists at the time, and his results were unknown until Maxwell published them in 1879.

Francis Ronalds delineated "intensity" (voltage) and "quantity" (current) for the dry pile - a high voltage source – in 1814 using a gold-leaf electrometer. He found for a dry pile that the relationship between the two parameters was not proportional under certain meteorological conditions.

Ohm's law in Georg Ohm's lab book.

Ohm did his work on resistance in the years 1825 and 1826, and published his results in 1827 as the book *Die galvanische Kette, mathematisch bearbeitet* ("The galvanic circuit investigated mathematically"). He drew considerable inspiration from Fourier's work on heat conduction in the theoretical explanation of his work. For experiments, he initially used voltaic piles, but later used a thermocouple as this provided a more stable voltage source in terms of internal resistance and constant voltage. He used a galvanometer to measure current, and knew that the voltage between the thermocouple terminals was proportional to the junction temperature. He then added test wires of varying length, diameter, and material to complete the circuit. He found that his data could be modeled through the equation

$$x = \frac{a}{b+l},$$

where x was the reading from the galvanometer, l was the length of the test conductor, a depended only on the thermocouple junction temperature, and b was a constant of the entire setup. From this, Ohm determined his law of proportionality and published his results.

Ohm's law was probably the most important of the early quantitative descriptions of the physics of electricity. We consider it almost obvious today. When Ohm first published his work, this was not the case; critics reacted to his treatment of the subject with hostility. They called his work a "web of naked fancies" and the German Minister of Education proclaimed that "a professor who preached such heresies was unworthy to teach science." The prevailing scientific philosophy in Germany at the time asserted that experiments need not be performed to develop an understanding of nature because nature is so well ordered, and that scientific truths may be deduced through reasoning alone. Also, Ohm's brother Martin, a mathematician, was battling the German educational system. These factors hindered the acceptance of Ohm's work, and his work did not become widely accepted until the 1840s. Fortunately, Ohm received recognition for his contributions to science well before he died.

In the 1850s, Ohm's law was known as such and was widely considered proved, and alternatives, such as "Barlow's law", were discredited, in terms of real applications to telegraph system design, as discussed by Samuel F. B. Morse in 1855.

While the old term for electrical conductance, the mho (the inverse of the resistance unit ohm), is still used, a new name, the siemens, was adopted in 1971, honoring Ernst Werner von Siemens. The siemens is preferred in formal papers.

In the 1920s, it was discovered that the current through a practical resistor actually has statistical fluctuations, which depend on temperature, even when voltage and resistance are exactly constant; this fluctuation, now known as Johnson–Nyquist noise, is due to the discrete nature of charge. This thermal effect implies that measurements of current and voltage that are taken over sufficiently short periods of time will yield ratios of V/I that fluctuate from the value of R implied by the time average or ensemble average of the measured current; Ohm's law remains correct for the average current, in the case of ordinary resistive materials.

Ohm's work long preceded Maxwell's equations and any understanding of frequency-dependent effects in AC circuits. Modern developments in electromagnetic theory and circuit theory do not contradict Ohm's law when they are evaluated within the appropriate limits.

Scope

Ohm's law is an empirical law, a generalization from many experiments that have shown that current is approximately proportional to electric field for most materials. It is less fundamental than Maxwell's equations and is not always obeyed. Any given material will break down under a strong-enough electric field, and some materials of interest in electrical engineering are "non-ohmic" under weak fields.

Ohm's law has been observed on a wide range of length scales. In the early 20th century, it was thought that Ohm's law would fail at the atomic scale, but experiments have not borne out this expectation. As of 2012, researchers have demonstrated that Ohm's law works for silicon wires as small as four atoms wide and one atom high.

Microscopic Origins

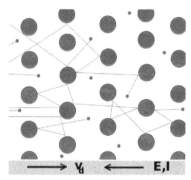

Drude Model electrons (shown here in blue) constantly bounce among heavier, stationary crystal ions (shown in red).

The dependence of the current density on the applied electric field is essentially quantum mechanical in nature. A qualitative description leading to Ohm's law can be based upon classical mechanics using the Drude model developed by Paul Drude in 1900.

The Drude model treats electrons (or other charge carriers) like pinballs bouncing among the ions that make up the structure of the material. Electrons will be accelerated in the opposite direction to the electric field by the average electric field at their location. With each collision, though, the electron is deflected in a random direction with a velocity that is much larger than the velocity gained by the electric field. The net result is that electrons take a zigzag path due to the collisions, but generally drift in a direction opposing the electric field.

The drift velocity then determines the electric current density and its relationship to E and is independent of the collisions. Drude calculated the average drift velocity from $p = -eE\tau$ where p is the average momentum, $-e$ is the charge of the electron and τ is the average time between the collisions. Since both the momentum and the current density are proportional to the drift velocity, the current density becomes proportional to the applied electric field; this leads to Ohm's law.

Hydraulic Analogy

A hydraulic analogy is sometimes used to describe Ohm's law. Water pressure, measured by pascals (or PSI), is the analog of voltage because establishing a water pressure difference between two points along a (horizontal) pipe causes water to flow. Water flow rate, as in liters per second, is the analog of current, as in coulombs per second. Finally, flow restrictors—such as apertures placed in pipes between points where the water pressure is measured—are the analog of resistors. We say that the rate of water flow through an aperture restrictor is proportional to the difference in water pressure across the restrictor. Similarly, the rate of flow of electrical charge, that is, the electric current, through an electrical resistor is proportional to the difference in voltage measured across the resistor.

Flow and pressure variables can be calculated in fluid flow network with the use of the hydraulic ohm analogy. The method can be applied to both steady and transient flow situations. In the linear laminar flow region, Poiseuille's law describes the hydraulic resistance of a pipe, but in the turbulent flow region the pressure–flow relations become nonlinear.

The hydraulic analogy to Ohm's law has been used, for example, to approximate blood flow through the circulatory system.

Circuit Analysis

Ohm's Law Triangle

Ohms law wheel with international unit symbols

In circuit analysis, three equivalent expressions of Ohm's law are used interchangeably:

$$I = \frac{V}{R} \quad \text{or} \quad V = IR \quad \text{or} \quad R = \frac{V}{I}.$$

Each equation is quoted by some sources as the defining relationship of Ohm's law, or all three are quoted, or derived from a proportional form, or even just the two that do not correspond to Ohm's original statement may sometimes be given.

The interchangeability of the equation may be represented by a triangle, where V (voltage) is placed on the top section, the I (current) is placed to the left section, and the R (resistance) is placed to the right. The line that divides the left and right sections indicate multiplication, and the divider between the top and bottom sections indicates division (hence the division bar).

Resistive Circuits

Resistors are circuit elements that impede the passage of electric charge in agreement with Ohm's law, and are designed to have a specific resistance value R. In a schematic diagram the resistor is shown as a zig-zag symbol. An element (resistor or conductor) that behaves according to Ohm's law over some operating range is referred to as an *ohmic device* (or an *ohmic resistor*) because Ohm's law and a single value for the resistance suffice to describe the behavior of the device over that range.

Ohm's law holds for circuits containing only resistive elements (no capacitances or inductances) for all forms of driving voltage or current, regardless of whether the driving voltage or current is constant (DC) or time-varying such as AC. At any instant of time Ohm's law is valid for such circuits.

Resistors which are in *series* or in *parallel* may be grouped together into a single "equivalent resistance" in order to apply Ohm's law in analyzing the circuit.

Reactive Circuits with Time-varying Signals

When reactive elements such as capacitors, inductors, or transmission lines are involved in a circuit to which AC or time-varying voltage or current is applied, the relationship between voltage and current becomes the solution to a differential equation, so Ohm's law (as defined above) does not directly apply since that form contains only resistances having value R, not complex impedances which may contain capacitance ("C") or inductance ("L").

Equations for time-invariant AC circuits take the same form as Ohm's law. However, the variables are generalized to complex numbers and the current and voltage waveforms are complex exponentials.

In this approach, a voltage or current waveform takes the form Ae^{st}, where t is time, s is a complex parameter, and A is a complex scalar. In any linear time-invariant system, all of the currents and voltages can be expressed with the same s parameter as the input to the system, allowing the time-varying complex exponential term to be canceled out and the system described algebraically in terms of the complex scalars in the current and voltage waveforms.

The complex generalization of resistance is impedance, usually denoted Z; it can be shown that for an inductor,

$$Z = sL$$

and for a capacitor,

$$Z = \frac{1}{sC}.$$

We can now write,

$$\mathbf{V} = \mathbf{I} \cdot \mathbf{Z}$$

where **V** and **I** are the complex scalars in the voltage and current respectively and **Z** is the complex impedance.

This form of Ohm's law, with Z taking the place of R, generalizes the simpler form. When Z is complex, only the real part is responsible for dissipating heat.

In the general AC circuit, Z varies strongly with the frequency parameter s, and so also will the relationship between voltage and current.

For the common case of a steady sinusoid, the s parameter is taken to be $j\omega$,, corresponding to a complex sinusoid $Ae^{j\omega}$.. The real parts of such complex current and voltage waveforms describe the actual sinusoidal currents and voltages in a circuit, which can be in different phases due to the different complex scalars.

Linear Approximations

Ohm's law is one of the basic equations used in the analysis of electrical circuits. It applies to both metal conductors and circuit components (resistors) specifically made for this behaviour. Both are ubiquitous in electrical engineering. Materials and components that obey Ohm's law are described as "ohmic" which means they produce the same value for resistance (R = V/I) regardless of the value of V or I which is applied and whether the applied voltage or current is DC (direct current) of either positive or negative polarity or AC (alternating current).

In a true ohmic device, the same value of resistance will be calculated from R = V/I regardless of the value of the applied voltage V. That is, the ratio of V/I is constant, and when current is plotted as a function of voltage the curve is *linear* (a straight line). If voltage is forced to some value V, then that voltage V divided by measured current I will equal R. Or if the current is forced to some value I, then the measured voltage V divided by that current I is also R. Since the plot of I versus V is a straight line, then it is also true that for any set of two different voltages V_1 and V_2 applied across a given device of resistance R, producing currents $I_1 = V_1/R$ and $I_2 = V_2/R$, that the ratio $(V_1-V_2)/(I_1-I_2)$ is also a constant equal to R. The operator "delta" (Δ) is used to represent a difference in a quantity, so we can write $\Delta V = V_1-V_2$ and $\Delta I = I_1-I_2$. Summarizing, for any truly ohmic device having resistance R, $V/I = \Delta V/\Delta I = R$ for any applied voltage or current or for the difference between any set of applied voltages or currents.

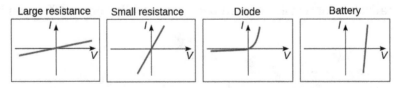

The I–V curves of four devices: Two resistors, a diode, and a battery. The two resistors follow Ohm's law: The plot is a straight line through the origin. The other two devices do *not* follow Ohm's law.

There are, however, components of electrical circuits which do not obey Ohm's law; that is, their relationship between current and voltage (their I–V curve) is *nonlinear* (or non-ohmic). An example is the p-n junction diode (curve at right). As seen in the figure, the current does not increase linearly with applied voltage for a diode. One can determine a value of current (I) for a given value of applied voltage (V) from the curve, but not from Ohm's law, since the value of "resistance" is not constant as a function of applied voltage. Further, the current only increases significantly if the applied voltage is positive, not negative. The ratio V/I for some point along the nonlinear curve is sometimes called the *static*, or *chordal*, or DC, resistance, but as seen in the figure the value of total V over total I varies depending on the particular point along the nonlinear curve which is chosen. This means the "DC resistance" V/I at some point on the curve is not the same as what would be determined by applying an AC signal having peak amplitude ΔV volts or ΔI amps centered at that same point along the curve and measuring $\Delta V/\Delta I$. However, in some diode applications, the AC signal applied to the device is small and it is possible to analyze the circuit in terms of the *dynamic*, *small-signal*, or *incremental* resistance, defined as the one over the slope of the V–I curve at the average value (DC operating point) of the voltage (that is, one over the derivative of current with respect to voltage). For sufficiently small signals, the dynamic resistance allows the Ohm's law small signal resistance to be calculated as approximately one over the slope of a line drawn tangentially to the V-I curve at the DC operating point.

Temperature Effects

Ohm's law has sometimes been stated as, "for a conductor in a given state, the electromotive force is proportional to the current produced." That is, that the resistance, the ratio of the applied electromotive force (or voltage) to the current, "does not vary with the current strength ." The qualifier "in a given state" is usually interpreted as meaning "at a constant temperature," since the resistivity of materials is usually temperature dependent. Because the conduction of current is related to Joule heating of the conducting body, according to Joule's first law, the temperature of a conducting body may change when it carries a current. The dependence of resistance on temperature therefore makes resistance depend upon the current in a typical experimental setup, making the law in this form difficult to directly verify. Maxwell and others worked out several methods to test the law experimentally in 1876, controlling for heating effects.

Relation to Heat Conductions

Ohm's principle predicts the flow of electrical charge (i.e. current) in electrical conductors when subjected to the influence of voltage differences; Jean-Baptiste-Joseph Fourier's principle predicts the flow of heat in heat conductors when subjected to the influence of temperature differences.

The same equation describes both phenomena, the equation's variables taking on different meanings in the two cases. Specifically, solving a heat conduction (Fourier) problem with *temperature* (the driving "force") and *flux of heat* (the rate of flow of the driven "quantity", i.e. heat energy) variables also solves an analogous electrical conduction (Ohm) problem having *electric potential* (the driving "force") and *electric current* (the rate of flow of the driven "quantity", i.e. charge) variables.

The basis of Fourier's work was his clear conception and definition of thermal conductivity. He assumed that, all else being the same, the flux of heat is strictly proportional to the gradient of temperature. Although undoubtedly true for small temperature gradients, strictly proportional behavior will be lost when real materials (e.g. ones having a thermal conductivity that is a function of temperature) are subjected to large temperature gradients.

A similar assumption is made in the statement of Ohm's law: other things being alike, the strength of the current at each point is proportional to the gradient of electric potential. The accuracy of the assumption that flow is proportional to the gradient is more readily tested, using modern measurement methods, for the electrical case than for the heat case.

Other Versions

Ohm's law, in the form above, is an extremely useful equation in the field of electrical/electronic engineering because it describes how voltage, current and resistance are interrelated on a "macroscopic" level, that is, commonly, as circuit elements in an electrical circuit. Physicists who study the electrical properties of matter at the microscopic level use a closely related and more general vector equation, sometimes also referred to as Ohm's law, having variables that are closely related to the V, I, and R scalar variables of Ohm's law, but which are each functions of position within the conductor. Physicists often use this continuum form of Ohm's Law:

$$\mathbf{E} = \rho \mathbf{J}$$

where "E" is the electric field vector with units of volts per meter (analogous to "V" of Ohm's law which has units of volts), "J" is the current density vector with units of amperes per unit area (analogous to "I" of Ohm's law which has units of amperes), and "ρ" is the resistivity with units of ohm·meters (analogous to "R" of Ohm's law which has units of ohms). The above equation is sometimes written as J = σ E where "σ" is the conductivity which is the reciprocal of ρ.

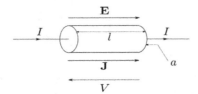

Current flowing through a uniform cylindrical conductor (such as a round wire) with a uniform field applied.

The voltage between two points is defined as:

$$\Delta V = -\int \mathbf{E} \cdot d\mathbf{l}$$

with $d\mathbf{l}$ the element of path along the integration of electric field vector E. If the applied E field is uniform and oriented along the length of the conductor as shown in the figure, then defining the voltage V in the usual convention of being opposite in direction to the field, and with the understanding that the voltage V is measured differentially across the length of the conductor allowing us to drop the Δ symbol, the above vector equation reduces to the scalar equation:

$$V = El \quad \text{or} \quad E = \frac{V}{l}.$$

Since the E field is uniform in the direction of wire length, for a conductor having uniformly consistent resistivity ρ, the current density J will also be uniform in any cross-sectional area and oriented in the direction of wire length, so we may write:

$$J = \frac{I}{a}.$$

Substituting the above 2 results (for E and J respectively) into the continuum form shown at the beginning of this section:

$$\frac{V}{l} = \frac{I}{a}\rho \quad \text{or} \quad V = I\rho\frac{l}{a}.$$

The electrical resistance of a uniform conductor is given in terms of resistivity by:

$$R = \rho\frac{l}{a}$$

where l is the length of the conductor in SI units of meters, a is the cross-sectional area (for a round wire $a = \pi r^2$ if r is radius) in units of meters squared, and ρ is the resistivity in units of ohm·meters.

After substitution of R from the above equation into the equation preceding it, the continuum form of Ohm's law for a uniform field (and uniform current density) oriented along the length of the conductor reduces to the more familiar form:

$$V = IR.$$

A perfect crystal lattice, with low enough thermal motion and no deviations from periodic structure, would have no resistivity, but a real metal has crystallographic defects, impurities, multiple isotopes, and thermal motion of the atoms. Electrons scatter from all of these, resulting in resistance to their flow.

The more complex generalized forms of Ohm's law are important to condensed matter physics, which studies the properties of matter and, in particular, its electronic structure. In broad terms, they fall under the topic of constitutive equations and the theory of transport coefficients.

Magnetic Effects

If an external B-field is present and the conductor is not at rest but moving at velocity v, then an extra term must be added to account for the current induced by the Lorentz force on the charge carriers.

$$\mathbf{J} = \sigma(\mathbf{E} + \mathbf{v} \times \mathbf{B})$$

In the rest frame of the moving conductor this term drops out because v= 0. There is no contradiction because the electric field in the rest frame differs from the E-field in the lab frame: E ' = E + v×B. Electric and magnetic fields are relative.

If the current J is alternating because the applied voltage or E-field varies in time, then reactance must be added to resistance to account for self-inductance. The reactance may be strong if the frequency is high or the conductor is coiled.

Conductive Fluids

In a conductive fluid, such as a plasma, there is a similar effect. Consider a fluid moving with the velocity $\vec{v}$ in a magnetic field $\vec{B}$. The relative motion induces an electric field $\vec{E}$ which exerts electric force on the charged particles giving rise to an electric current $\vec{J}$. The equation of motion for the electron gas, with a number density n_e, is written as

$$m_e n_e \frac{d\vec{v}_e}{dt} = -n_e e\vec{E} + n_e m_e \nu(v_i - v_e) - en_e \vec{v}_e \times \vec{B},$$

where e, m_e and $\vec{v}_e$ are the charge, mass and velocity of the electrons, respectively. Also, ν is the frequency of collisions of the electrons with ions which have a velocity field $\vec{v}_i$. Since, the electron has a very small mass compared with that of ions, we can ignore the left hand side of the above equation to write

$$\sigma(\vec{E} + \vec{v} \times \vec{B}) = \vec{J},$$

where we have used the definition of the current density, and also put $\sigma = \frac{n_e e^2}{\nu m_e}$ which is the electrical conductivity. This equation can also be equivalently written as

$$\vec{E} + \vec{v} \times \vec{B} = \rho \vec{J},$$

where $\rho = \sigma^{-1}$ is the electrical resistivity. It is also common to write η instead of ρ which can be confusing since it is the same notation used for the magnetic diffusivity defined as $\eta = 1/\mu_0 \sigma$.

References

- Heinz E Knoepfel (2000). Magnetic Fields: A comprehensive theoretical treatise for practical use. Wiley. p. 4. ISBN 0-471-32205-9.

- George E. Owen (2003). Electromagnetic Theory (Reprint of 1963 ed.). Courier-Dover Publications. p. 213. ISBN 0-486-42830-3.

- David J Griffiths (1999). Introduction to Electrodynamics (3rd ed.). Pearson/Addison-Wesley. pp. 322–323. ISBN 0-13-805326-X.

- J.C. Slater and N.H. Frank (1969). Electromagnetism (Reprint of 1947 ed.). Courier Dover Publications. p. 83. ISBN 0-486-62263-0.

- Daniel M. Siegel (2003). Innovation in Maxwell's Electromagnetic Theory: Molecular Vortices, Displacement Current, and Light. Cambridge University Press. pp. 96–98. ISBN 0-521-53329-5.

- Hall. p. 323. ISBN 0-13-805326-X. and Tai L. Chow (2006). Introduction to Electromagnetic Theory. Jones & Bartlett. p. 204. ISBN 0-7637-3827-1.

- Ralph Morrison, Grounding and Shielding Techniques in Instrumentation Wiley-Interscience (1986) ISBN 0471838055

- Serway, Raymond A.; Jewett, John W. (2004). Physics for Scientists and Engineers (6th ed.). Brooks/Cole. ISBN 0-534-40842-7.

- Tipler, Paul (2004). Physics for Scientists and Engineers: Electricity, Magnetism, Light, and Elementary Modern Physics (5th ed.). W. H. Freeman. ISBN 0-7167-0810-8.

- Graham, Howard Johnson, Martin (2002). High-speed signal propagation : advanced black magic (10. printing. ed.). Upper Saddle River, NJ: Prentice Hall PTR. ISBN 0-13-084408-X.

- Sadiku, M. N. O. (2007). Elements of Electromagnetics (fourth ed.). New York (USA)/Oxford (UK): Oxford University Press. p. 386. ISBN 0-19-530048-3.

- Ulaby, Fawwaz (2007). Fundamentals of applied electromagnetics (5th ed.). Pearson:Prentice Hall. p. 255. ISBN 0-13-241326-4.

- Faraday, Michael; Day, P. (1999-02-01). The philosopher's tree: a selection of Michael Faraday's writings. CRC Press. p. 71. ISBN 978-0-7503-0570-9. Retrieved 28 August 2011.

- Griffiths, David J. (1999). Introduction to Electrodynamics (Third ed.). Upper Saddle River NJ: Prentice Hall. pp. 301–303. ISBN 0-13-805326-X.

- Roger F Harrington (2003). Introduction to electromagnetic engineering. Mineola, NY: Dover Publications. p. 56. ISBN 0-486-43241-6.

- Tipler, Paul A.; Mosca, Gene (2008). Physics for Scientists and Engineers (6th ed.). New York: W. H. Freeman and Company. ISBN 0-7167-8964-7. LCCN 2007010418.

- Young, Hugh D.; Freedman, Roger A. (2010). Sears and Zemansky's University Physics : With Modern Physics (13th ed.). Addison-Wesley (Pearson). ISBN 978-0-321-69686-1.

- I.S. Grant, W.R. Phillips (2008). Electromagnetism (2nd ed.). Manchester Physics, John Wiley & Sons. ISBN 978-0-471-92712-9.

- Sanford P. Bordeau (1982) Volts to Hertz...the Rise of Electricity. Burgess Publishing Company, Minneapolis, MN. pp.86–107, ISBN 0-8087-4908-0

- Ronalds, B.F. (2016). Sir Francis Ronalds: Father of the Electric Telegraph. London: Imperial College Press. ISBN 978-1-78326-917-4.

- Herbert Schnädelbach, Philosophy in Germany 1831-1933, pages 78-79, Cambridge University Press, 1984 ISBN 0521296463.

- Purcell, Edward M. (1985), Electricity and magnetism, Berkeley Physics Course, 2 (2nd ed.), McGraw-Hill, p. 129, ISBN 0-07-004908-4

- A. Akers; M. Gassman & R. Smith (2006). Hydraulic Power System Analysis. New York: Taylor & Francis. Chapter 13. ISBN 0-8247-9956-9.

- Alvin M. Halpern & Erich Erlbach (1998). Schaum's outline of theory and problems of beginning physics II. McGraw-Hill Professional. p. 140. ISBN 978-0-07-025707-8.

- Kenneth L. Kaiser (2004). Electromagnetic Compatibility Handbook. CRC Press. pp. 13–52. ISBN 978-0-8493-2087-3.

- Horowitz, Paul; Winfield Hill (1989). The Art of Electronics (2nd ed.). Cambridge University Press. p. 13. ISBN 0-521-37095-7.

Understanding Electric Power and Current

Electric power is the degree of energy being transmitted by electric circuits. Electric current is the movement of electric charge. Other topics covered in this chapter are electric machine, passive sign convention, AC power, power station etc. This section has been carefully written to provide an easy understanding of electric power and current.

Electric Power

Electric power is transmitted on overhead lines like these, and also on underground high voltage cables.

Electric power is the rate at which electrical energy is transferred by an electric circuit. The SI unit of power is the watt, one joule per second.

Electric power is usually produced by electric generators, but can also be supplied by sources such as electric batteries. It is usually supplied to businesses and homes by the electric power industry through an electric power grid. Electric power is usually sold by the kilowatt hour (3.6 MJ) which is the product of power in kilowatts multiplied by running time in hours. Electric utilities measure power using an electricity meter, which keeps a running total of the electric energy delivered to a customer.

Electrical power provides a low entropy form of energy and can be converted into motion or other forms of energy with high efficiency.

Definition

Electric power, like mechanical power, is the rate of doing work, measured in watts, and represented by the letter P. The term *wattage* is used colloquially to mean "electric power in watts." The electric power in watts produced by an electric current I consisting of a charge of Q coulombs every t seconds passing through an electric potential (voltage) difference of V is

$$P = \text{work done per unit time} = \frac{VQ}{t} = VI$$

where

Q is electric charge in coulombs

t is time in seconds

I is electric current in amperes

V is electric potential or voltage in volts

Explanation

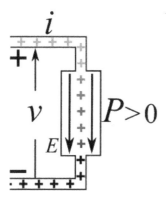

Animation showing electric load

Electric power is transformed to other forms of energy when electric charges move through an electric potential (voltage) difference, which occurs in electrical components in electric circuits. From the standpoint of electric power, components in an electric circuit can be divided into two categories:

Passive devices or loads: When electric charges move through a potential difference from a higher to a lower voltage, that is when conventional current (positive charge) moves from the positive (+) terminal to the negative (−) terminal, work is done by the charges on the device. The potential energy of the charges due to the voltage between the terminals is converted to kinetic energy in the device. These devices are called *passive* components or *loads*; they 'consume' electric power from the circuit, converting it to other forms of energy such as mechanical work, heat, light, etc. Examples are electrical appliances, such as light bulbs, electric motors, and electric heaters. In alternating current (AC) circuits the direction of the voltage periodically reverses, but the current always flows from the higher potential to the lower potential side.

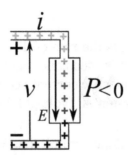

Animation showing power source

Active devices or power sources: If the charges are moved by an 'exterior force' through the device in the direction from the lower electric potential to the higher, (so positive charge moves from the negative to the positive terminal), work will be done *on* the charges, and energy is being converted to electric potential energy from some other type of energy, such as mechanical energy or chemical energy. Devices in which this occurs are called *active* devices or *power sources*; such as electric generators and batteries.

Some devices can be either a source or a load, depending on the voltage and current through them. For example, a rechargeable battery acts as a source when it provides power to a circuit, but as a load when it is connected to a battery charger and is being recharged.

Passive Sign Convention

Since electric power can flow either into or out of a component, a convention is needed for which direction represents positive power flow. Electric power flowing *out* of a circuit *into* a component is arbitrarily defined to have a positive sign, while power flowing *into* a circuit from a component is defined to have a negative sign. Thus passive components have positive power consumption, while power sources have negative power consumption. This is called the *passive sign convention*.

Resistive Circuits

In the case of resistive (Ohmic, or linear) loads, Joule's law can be combined with Ohm's law ($V = I \cdot R$) to produce alternative expressions for the amount of power that is dissipated:

$$P = IV = I^2 R = \frac{V^2}{R},$$

where R is the electrical resistance.

Alternating Current

In alternating current circuits, energy storage elements such as inductance and capacitance may result in periodic reversals of the direction of energy flow. The portion of power flow that, averaged over a complete cycle of the AC waveform, results in net transfer of energy in one direction is known as real power (also referred to as active power). That portion of power flow due to stored energy, that returns to the source in each cycle, is known as reactive power. The real power P in watts consumed by a device is given by

$$P = \frac{1}{2}V_p I_p \cos\theta = V_{rms}I_{rms}\cos\theta$$

where

V_p is the peak voltage in volts

I_p is the peak current in amperes

V_{rms} is the root-mean-square voltage in volts

I_{rms} is the root-mean-square current in amperes

θ is the phase angle between the current and voltage sine waves

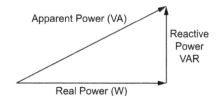

Power triangle: The components of AC power

The relationship between real power, reactive power and apparent power can be expressed by representing the quantities as vectors. Real power is represented as a horizontal vector and reactive power is represented as a vertical vector. The apparent power vector is the hypotenuse of a right triangle formed by connecting the real and reactive power vectors. This representation is often called the *power triangle*. Using the Pythagorean Theorem, the relationship among real, reactive and apparent power is:

$$(\text{apparent power})^2 = (\text{real power})^2 + (\text{reactive power})^2$$

Real and reactive powers can also be calculated directly from the apparent power, when the current and voltage are both sinusoids with a known phase angle θ between them:

$$(\text{real power}) = (\text{apparent power})\cos\theta$$

$$(\text{reactive power}) = (\text{apparent power})\sin\theta$$

The ratio of real power to apparent power is called power factor and is a number always between 0 and 1. Where the currents and voltages have non-sinusoidal forms, power factor is generalized to include the effects of distortion

Electromagnetic Fields

Electrical energy flows wherever electric and magnetic fields exist together and fluctuate in the same place. The simplest example of this is in electrical circuits, as the preceding section showed. In the general case, however, the simple equation $P = IV$ must be replaced by a more complex calculation, the integral of the cross-product of the electrical and magnetic field vectors over a specified area, thus:

$$P = \int_S (\mathbf{E} \times \mathbf{H}) \cdot d\mathbf{A}.$$

The result is a scalar since it is the *surface integral* of the *Poynting vector*.

Generation

The fundamental principles of much electricity generation were discovered during the 1820s and early 1830s by the British scientist Michael Faraday. His basic method is still used today: electricity is generated by the movement of a loop of wire, or disc of copper between the poles of a magnet.

For electric utilities, it is the first process in the delivery of electricity to consumers. The other processes, electricity transmission, distribution, and electrical power storage and recovery using pumped-storage methods are normally carried out by the electric power industry.

Electricity is mostly generated at a power station by electromechanical generators, driven by heat engines heated by combustion, geothermal power or nuclear fission. Other generators are driven by the kinetic energy of flowing water and wind. There are many other technologies that are used to generate electricity such as solar photovoltaics.

A battery is a device consisting of one or more electrochemical cells that convert stored chemical energy into electrical energy. Since the invention of the first battery (or "voltaic pile") in 1800 by Alessandro Volta and especially since the technically improved Daniell cell in 1836, batteries have become a common power source for many household and industrial applications. According to a 2005 estimate, the worldwide battery industry generates US$48 billion in sales each year, with 6% annual growth. There are two types of batteries: primary batteries (disposable batteries), which are designed to be used once and discarded, and secondary batteries (rechargeable batteries), which are designed to be recharged and used multiple times. Batteries come in many sizes, from miniature cells used to power hearing aids and wristwatches to battery banks the size of rooms that provide standby power for telephone exchanges and computer data centers.

Electric Power Industry

The electric power industry provides the production and delivery of power, in sufficient quantities to areas that need electricity, through a grid connection. The grid distributes electrical energy to customers. Electric power is generated by central power stations or by distributed generation.

Many households and businesses need access to electricity, especially in developed nations, the demand being scarcer in developing nations. Demand for electricity is derived from the requirement for electricity in order to operate domestic appliances, office equipment, industrial machinery and provide sufficient energy for both domestic and commercial lighting, heating, cooking and industrial processes. Because of this aspect the industry is viewed as part of the public utility infrastructure.

Electric Current

An electric current is a flow of electric charge. In electric circuits this charge is often carried by moving electrons in a wire. It can also be carried by ions in an electrolyte, or by both ions and electrons such as in a plasma.

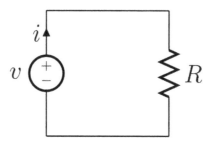

A simple electric circuit, where current is represented by the letter *i*. The relationship between the voltage (V), resistance (R), and current (I) is V=IR; this is known as Ohm's Law.

The SI unit for measuring an electric current is the ampere, which is the flow of electric charge across a surface at the rate of one coulomb per second. Electric current is measured using a device called an ammeter.

Electric currents cause Joule heating, which creates light in incandescent light bulbs. They also create magnetic fields, which are used in motors, inductors and generators.

The particles that carry the charge in an electric current are called charge carriers. In metals, one or more electrons from each atom are loosely bound to the atom, and can move freely about within the metal. These conduction electrons are the charge carriers in metal conductors.

Symbol

The conventional symbol for current is *I*, which originates from the French phrase *intensité de courant*, meaning *current intensity*. Current intensity is often referred to simply as *current*. The *I* symbol was used by André-Marie Ampère, after whom the unit of electric current is named, in formulating the eponymous Ampère's force law, which he discovered in 1820. The notation travelled from France to Great Britain, where it became standard, although at least one journal did not change from using *C* to *I* until 1896.

Conventions

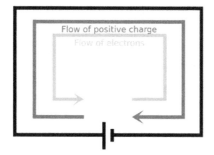

The electrons, the charge carriers in an electrical circuit, flow in the opposite direction of the conventional electric current.

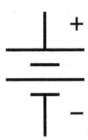

The symbol for a battery in a circuit diagram.

In a conductive material, the moving charged particles which constitute the electric current are called charge carriers. In metals, which make up the wires and other conductors in most electrical circuits, the positively charged atomic nuclei are held in a fixed position, and the negatively charged electrons are free to move, carrying their charge from one place to another. In other materials, notably the semiconductors, the charge carriers can be positive *or* negative, depending on the dopant used. Positive and negative charge carriers may even be present at the same time, as happens in an electrochemical cell.

A flow of positive charges gives the same electric current, and has the same effect in a circuit, as an equal flow of negative charges in the opposite direction. Since current can be the flow of either positive or negative charges, or both, a convention is needed for the direction of current that is independent of the type of charge carriers. The direction of *conventional current* is arbitrarily defined as the same direction as positive charges flow.

The consequence of this convention is that electrons, the charge carriers in metal wires and most other parts of electric circuits, flow in the opposite direction of conventional current flow in an electrical circuit.

Reference Direction

Since the current in a wire or component can flow in either direction, when a variable I is defined to represent that current, the direction representing positive current must be specified, usually by an arrow on the circuit schematic diagram. This is called the *reference direction* of current I. If the current flows in the opposite direction, the variable I has a negative value.

When analyzing electrical circuits, the actual direction of current through a specific circuit element is usually unknown. Consequently, the reference directions of currents are often assigned arbitrarily. When the circuit is solved, a negative value for the variable means that the actual direction of current through that circuit element is opposite that of the chosen reference direction. In electronic circuits, the reference current directions are often chosen so that all currents are toward ground. This often corresponds to the actual current direction, because in many circuits the power supply voltage is positive with respect to ground.

Ohm's Law

Ohm's law states that the current through a conductor between two points is directly proportional to the potential difference across the two points. Introducing the constant of proportionality, the resistance, one arrives at the usual mathematical equation that describes this relationship:

$$I = \frac{V}{R}$$

where I is the current through the conductor in units of amperes, V is the potential difference measured *across* the conductor in units of volts, and R is the resistance of the conductor in units of ohms. More specifically, Ohm's law states that the R in this relation is constant, independent of the current.

AC and DC

The abbreviations *AC* and *DC* are often used to mean simply *alternating* and *direct*, as when they modify *current* or *voltage*.

Direct Current

Direct current (DC) is the unidirectional flow of electric charge. Direct current is produced by sources such as batteries, thermocouples, solar cells, and commutator-type electric machines of the dynamo type. Direct current may flow in a conductor such as a wire, but can also flow through semiconductors, insulators, or even through a vacuum as in electron or ion beams. The electric charge flows in a constant direction, distinguishing it from alternating current (AC). A term formerly used for *direct current* was galvanic current.

Alternating Current

In alternating current (AC, also ac), the movement of electric charge periodically reverses direction. In direct current (DC, also dc), the flow of electric charge is only in one direction.

AC is the form of electric power delivered to businesses and residences. The usual waveform of an AC power circuit is a sine wave. Certain applications use different waveforms, such as triangular or square waves. Audio and radio signals carried on electrical wires are also examples of alternating current. An important goal in these applications is recovery of information encoded (or *modulated*) onto the AC signal.

Occurrences

Natural observable examples of electrical current include lightning, static electricity, and the solar wind, the source of the polar auroras.

Man-made occurrences of electric current include the flow of conduction electrons in metal wires such as the overhead power lines that deliver electrical energy across long distances and the smaller wires within electrical and electronic equipment. Eddy currents are electric currents that occur in conductors exposed to changing magnetic fields. Similarly, electric currents occur, particularly in the surface, of conductors exposed to electromagnetic waves. When oscillating electric currents flow at the correct voltages within radio antennas, radio waves are generated.

In electronics, other forms of electric current include the flow of electrons through resistors or through the vacuum in a vacuum tube, the flow of ions inside a battery or a neuron, and the flow of holes within a semiconductor.

Current Measurement

Current can be measured using an ammeter.

At the circuit level, there are various techniques that can be used to measure current:

- Shunt resistors
- Hall effect current sensor transducers
- Transformers (however DC cannot be measured)
- Magnetoresistive field sensors

Resistive Heating

Joule heating, also known as *ohmic heating* and *resistive heating*, is the process by which the passage of an electric current through a conductor releases heat. It was first studied by James Prescott Joule in 1841. Joule immersed a length of wire in a fixed mass of water and measured the temperature rise due to a known current through the wire for a 30 minute period. By varying the current and the length of the wire he deduced that the heat produced was proportional to the square of the current multiplied by the electrical resistance of the wire.

$$Q \propto I \ R$$

This relationship is known as Joule's First Law. The SI unit of energy was subsequently named the joule and given the symbol J. The commonly known unit of power, the watt, is equivalent to one joule per second.

Electromagnetism

Electromagnet

In an electromagnet a coil, of a large number of circular turns of insulated wire, wrapped on a cylindrical core, behaves like a magnet when an electric current flows through it. When the current is switched off, the coil loses its magnetism immediately. We call such a device as an electromagnet.

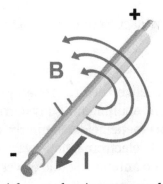

According to Ampère's law, an electric current produces a magnetic field.

Electric current produces a magnetic field. The magnetic field can be visualized as a pattern of circular field lines surrounding the wire that persists as long as there is current.

Magnetism can also produce electric currents. When a changing magnetic field is applied to a conductor, an Electromotive force (EMF) is produced, and when there is a suitable path, this causes current.

Electric current can be directly measured with a galvanometer, but this method involves breaking the electrical circuit, which is sometimes inconvenient. Current can also be measured without breaking the circuit by detecting the magnetic field associated with the current. Devices used for this include Hall effect sensors, current clamps, current transformers, and Rogowski coils.

Radio Waves

When an electric current flows in a suitably shaped conductor at radio frequencies radio waves can be generated. These travel at the speed of light and can cause electric currents in distant conductors.

Conduction Mechanisms in Various Media

In metallic solids, electric charge flows by means of electrons, from lower to higher electrical potential. In other media, any stream of charged objects (ions, for example) may constitute an electric current. To provide a definition of current independent of the type of charge carriers, *conventional current* is defined as moving in the same direction as the positive charge flow. So, in metals where the charge carriers (electrons) are negative, conventional current is in the opposite direction as the electrons. In conductors where the charge carriers are positive, conventional current is in the same direction as the charge carriers.

In a vacuum, a beam of ions or electrons may be formed. In other conductive materials, the electric current is due to the flow of both positively and negatively charged particles at the same time. In still others, the current is entirely due to positive charge flow. For example, the electric currents in electrolytes are flows of positively and negatively charged ions. In a common lead-acid electrochemical cell, electric currents are composed of positive hydrogen ions (protons) flowing in one direction, and negative sulfate ions flowing in the other. Electric currents in sparks or plasma are flows of electrons as well as positive and negative ions. In ice and in certain solid electrolytes, the electric current is entirely composed of flowing ions.

Metals

In a metal, some of the outer electrons in each atom are not bound to the individual atom as they are in insulating materials, but are free to move within the metal lattice. These conduction electrons can serve as charge carriers, carrying a current. Metals are particularly conductive because there are a large number of these free electrons, typically one per atom in the lattice. With no external electric field applied, these electrons move about randomly due to thermal energy but, on average, there is zero net current within the metal. At room temperature, the average speed of these random motions is 10^6 metres per second. Given a surface through which a metal wire passes, electrons move in both directions across the surface at an equal rate. As George Gamow wrote in his popular science book, *One, Two, Three...Infinity* (1947), "The metallic substances differ from all other materials by the fact that the outer shells of their atoms are bound rather loosely, and often let one of their electrons go free. Thus the interior of a metal is filled up with a large number of unattached electrons that travel aimlessly around like a crowd of displaced persons. When a metal wire is subjected to electric force applied on its opposite ends, these free electrons rush in the direction of the force, thus forming what we call an electric current."

When a metal wire is connected across the two terminals of a DC voltage source such as a battery, the source places an electric field across the conductor. The moment contact is made, the free electrons of the conductor are forced to drift toward the positive terminal under the influence of this field. The free electrons are therefore the charge carrier in a typical solid conductor.

For a steady flow of charge through a surface, the current I (in amperes) can be calculated with the following equation:

$$I = \frac{Q}{t},$$

where Q is the electric charge transferred through the surface over a time t. If Q and t are measured in coulombs and seconds respectively, I is in amperes.

More generally, electric current can be represented as the rate at which charge flows through a given surface as:

$$I = \frac{dQ}{dt}.$$

Electrolytes

Electric currents in electrolytes are flows of electrically charged particles (ions). For example, if an electric field is placed across a solution of Na^+ and Cl^- (and conditions are right) the sodium ions move towards the negative electrode (cathode), while the chloride ions move towards the positive electrode (anode). Reactions take place at both electrode surfaces, absorbing each ion.

Water-ice and certain solid electrolytes called proton conductors contain positive hydrogen ions ("protons") that are mobile. In these materials, electric currents are composed of moving protons, as opposed to the moving electrons in metals.

In certain electrolyte mixtures, brightly coloured ions are the moving electric charges. The slow progress of the colour makes the current visible.

Gases and Plasmas

In air and other ordinary gases below the breakdown field, the dominant source of electrical conduction is via relatively few mobile ions produced by radioactive gases, ultraviolet light, or cosmic rays. Since the electrical conductivity is low, gases are dielectrics or insulators. However, once the applied electric field approaches the breakdown value, free electrons become sufficiently accelerated by the electric field to create additional free electrons by colliding, and ionizing, neutral gas atoms or molecules in a process called avalanche breakdown. The breakdown process forms a plasma that contains enough mobile electrons and positive ions to make it an electrical conductor. In the process, it forms a light emitting conductive path, such as a spark, arc or lightning.

Plasma is the state of matter where some of the electrons in a gas are stripped or "ionized" from their molecules or atoms. A plasma can be formed by high temperature, or by application of a high electric or alternating magnetic field as noted above. Due to their lower mass, the electrons in a plasma accelerate more quickly in response to an electric field than the heavier positive ions, and hence carry the bulk of the current. The free ions recombine to create new chemical compounds (for example, breaking atmospheric oxygen into single oxygen [$O_2 \rightarrow 2O$], which then recombine creating ozone [O_3]).

Vacuum

Since a "perfect vacuum" contains no charged particles, it normally behaves as a perfect insulator. However, metal electrode surfaces can cause a region of the vacuum to become conductive by injecting free electrons or ions through either field electron emission or thermionic emission. Thermionic emission occurs when the thermal energy exceeds the metal's work function, while field electron emission occurs when the electric field at the surface of the metal is high enough to cause tunneling, which results in the ejection of free electrons from the metal into the vacuum. Externally heated electrodes are often used to generate an electron cloud as in the filament or indirectly heated cathode of vacuum tubes. Cold electrodes can also spontaneously produce electron clouds via thermionic emission when small incandescent regions (called cathode spots or anode spots) are formed. These are incandescent regions of the electrode surface that are created by a localized high current. These regions may be initiated by field electron emission, but are then sustained by localized thermionic emission once a vacuum arc forms. These small electron-emitting regions can form quite rapidly, even explosively, on a metal surface subjected to a high electrical field. Vacuum tubes and sprytrons are some of the electronic switching and amplifying devices based on vacuum conductivity.

Superconductivity

Superconductivity is a phenomenon of exactly zero electrical resistance and expulsion of magnetic fields occurring in certain materials when cooled below a characteristic critical temperature. It was discovered by Heike Kamerlingh Onnes on April 8, 1911 in Leiden. Like ferromagnetism and atomic spectral lines, superconductivity is a quantum mechanical phenomenon. It is characterized by the Meissner effect, the complete ejection of magnetic field lines from the interior of the superconductor as it transitions into the superconducting state. The occurrence of the Meissner effect indicates that superconductivity cannot be understood simply as the idealization of *perfect conductivity* in classical physics.

Semiconductor

In a semiconductor it is sometimes useful to think of the current as due to the flow of positive "holes" (the mobile positive charge carriers that are places where the semiconductor crystal is missing a valence electron). This is the case in a p-type semiconductor. A semiconductor has electrical conductivity intermediate in magnitude between that of a conductor and an insulator. This means a conductivity roughly in the range of 10^{-2} to 10^4 siemens per centimeter (S·cm^{-1}).

In the classic crystalline semiconductors, electrons can have energies only within certain bands (i.e. ranges of levels of energy). Energetically, these bands are located between the energy of the ground state, the state in which electrons are tightly bound to the atomic nuclei of the material, and the free electron energy, the latter describing the energy required for an electron to escape entirely from the material. The energy bands each correspond to a large number of discrete quantum states of the electrons, and most of the states with low energy (closer to the nucleus) are occupied, up to a particular band called the *valence band*. Semiconductors and insulators are distinguished from metals because the valence band in any given metal is nearly filled with electrons under usual operating conditions, while very few (semiconductor) or virtually none (insulator) of them are available in the *conduction band*, the band immediately above the valence band.

The ease of exciting electrons in the semiconductor from the valence band to the conduction band depends on the band gap between the bands. The size of this energy band gap serves as an arbitrary dividing line (roughly 4 eV) between semiconductors and insulators.

With covalent bonds, an electron moves by hopping to a neighboring bond. The Pauli exclusion principle requires that the electron be lifted into the higher anti-bonding state of that bond. For delocalized states, for example in one dimension – that is in a nanowire, for every energy there is a state with electrons flowing in one direction and another state with the electrons flowing in the other. For a net current to flow, more states for one direction than for the other direction must be occupied. For this to occur, energy is required, as in the semiconductor the next higher states lie above the band gap. Often this is stated as: full bands do not contribute to the electrical conductivity. However, as a semiconductor's temperature rises above absolute zero, there is more energy in the semiconductor to spend on lattice vibration and on exciting electrons into the conduction band. The current-carrying electrons in the conduction band are known as *free electrons*, though they are often simply called *electrons* if that is clear in context.

Current Density and Ohm's Law

Current density is a measure of the density of an electric current. It is defined as a vector whose magnitude is the electric current per cross-sectional area. In SI units, the current density is measured in amperes per square metre.

$$I = \int \vec{J} \cdot d\vec{A}$$

where I is current in the conductor, $\vec{J}$ is the current density, and $d\vec{A}$ is the differential cross-sectional area vector.

The current density (current per unit area) $\vec{J}$ in materials with finite resistance is directly proportional to the electric field $\vec{E}$ in the medium. The proportionality constant is called the conductivity σ of the material, whose value depends on the material concerned and, in general, is dependent on the temperature of the material:

$$\vec{J} = \sigma \vec{E}$$

The reciprocal of the conductivity σ of the material is called the resistivity ρ of the material and the above equation, when written in terms of resistivity becomes:

$$\vec{J} = \frac{\vec{E}}{\rho} \text{ or}$$

$$\vec{E} = \rho\vec{J}$$

Conduction in semiconductor devices may occur by a combination of drift and diffusion, which is proportional to diffusion constant D and charge density α_q. The current density is then:

$$J = \sigma E + Dq\nabla n,$$

with q being the elementary charge and n the electron density. The carriers move in the direction of decreasing concentration, so for electrons a positive current results for a positive density gradient. If the carriers are holes, replace electron density n by the negative of the hole density p.

In linear anisotropic materials, σ, ρ and D are tensors.

In linear materials such as metals, and under low frequencies, the current density across the conductor surface is uniform. In such conditions, Ohm's law states that the current is directly proportional to the potential difference between two ends (across) of that metal (ideal) resistor (or other ohmic device):

$$I = \frac{V}{R},$$

where I is the current, measured in amperes; V is the potential difference, measured in volts; and R is the resistance, measured in ohms. For alternating currents, especially at higher frequencies, skin effect causes the current to spread unevenly across the conductor cross-section, with higher density near the surface, thus increasing the apparent resistance.

Drift Speed

The mobile charged particles within a conductor move constantly in random directions, like the particles of a gas. To create a net flow of charge, the particles must also move together with an average drift rate. Electrons are the charge carriers in metals and they follow an erratic path, bouncing from atom to atom, but generally drifting in the opposite direction of the electric field. The speed they drift at can be calculated from the equation:

$$I = nAvQ,$$

where

I is the electric current

n is number of charged particles per unit volume (or charge carrier density)

A is the cross-sectional area of the conductor

v is the drift velocity, and

Q is the charge on each particle.

Typically, electric charges in solids flow slowly. For example, in a copper wire of cross-section

0.5 mm², carrying a current of 5 A, the drift velocity of the electrons is on the order of a millimetre per second. To take a different example, in the near-vacuum inside a cathode ray tube, the electrons travel in near-straight lines at about a tenth of the speed of light.

Any accelerating electric charge, and therefore any changing electric current, gives rise to an electromagnetic wave that propagates at very high speed outside the surface of the conductor. This speed is usually a significant fraction of the speed of light, as can be deduced from Maxwell's Equations, and is therefore many times faster than the drift velocity of the electrons. For example, in AC power lines, the waves of electromagnetic energy propagate through the space between the wires, moving from a source to a distant load, even though the electrons in the wires only move back and forth over a tiny distance.

The ratio of the speed of the electromagnetic wave to the speed of light in free space is called the velocity factor, and depends on the electromagnetic properties of the conductor and the insulating materials surrounding it, and on their shape and size.

The magnitudes (but, not the natures) of these three velocities can be illustrated by an analogy with the three similar velocities associated with gases.

- The low drift velocity of charge carriers is analogous to air motion; in other words, winds.

- The high speed of electromagnetic waves is roughly analogous to the speed of sound in a gas (these waves move through the medium much faster than any individual particles do)

- The random motion of charges is analogous to heat – the thermal velocity of randomly vibrating gas particles.

Alternating Current

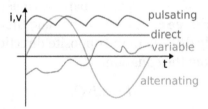

Alternating current (green curve). The horizontal axis measures time; the vertical, current or voltage.

Alternating current (AC), is an electric current in which the flow of electric charge periodically reverses direction, whereas in direct current (DC, also dc), the flow of electric charge is only in one direction. AC is the form in which electric power is delivered to businesses and residences, and it is the form of electric power that consumers typically use when they plug kitchen appliances, televisions and electric lamps into a wall socket. A common example of DC power is a battery cell in a flashlight. The abbreviations *AC* and *DC* are often used to mean simply *alternating* and *direct*, as when they modify *current* or *voltage*.

The usual waveform of alternating current in most electric power circuits is a sine wave. In certain applications, different waveforms are used, such as triangular or square waves. Audio and radio signals carried on electrical wires are also examples of alternating current. These types of alternating current carry information encoded (or modulated) onto the AC signal, such as sound (audio) or images (video). These currents typically alternate at higher frequencies than those used in power transmission.

Transmission, Distribution, and Domestic Power Supply

Electric power is distributed as alternating current because AC voltage may be increased or decreased with a transformer. This allows the power to be transmitted through power lines efficiently at high voltage, which reduces the power lost as heat due to resistance of the wire, and transformed to a lower, safer, voltage for use. Use of a higher voltage leads to significantly more efficient transmission of power. The power losses (P_L) in a conductor are a product of the square of the current (I) and the resistance (R) of the conductor, described by the formula

$$P_L = I^2 R.$$

This means that when transmitting a fixed power on a given wire, if the current is halved (i.e. the voltage is doubled), the power loss will be four times less.

The power transmitted is equal to the product of the current and the voltage (assuming no phase difference); that is,

$$P_T = IV.$$

Consequently, power transmitted at a higher voltage requires less loss-producing current than for the same power at a lower voltage. Power is often transmitted at hundreds of kilovolts, and transformed to 100–240 volts for domestic use.

High voltage transmission lines deliver power from electric generation plants over long distances using alternating current. These lines are located in eastern Utah.

High voltages have disadvantages, the main one being the increased insulation required, and generally increased difficulty in their safe handling. In a power plant, power is generated at a convenient voltage for the design of a generator, and then stepped up to a high voltage for transmission. Near the loads, the transmission voltage is stepped down to the voltages used by equipment. Consumer voltages vary depending on the country and size of load, but generally motors and lighting are built to use up to a few hundred volts between phases. The voltage delivered to equipment such as lighting and motor loads is standardized, with an allowable range of voltage over which equipment is expected to operate. Standard power utilization voltages and percentage tolerance vary in the different mains power systems found in the world. High-voltage direct-current (HVDC) electric power transmission systems have become more viable as technology has provided efficient means of changing the voltage of DC power. HVDC systems, however, tend to be more expensive and less efficient over shorter distances than transformers. Transmission with high voltage direct current was not feasible in the early days of electric power transmission, as there was then no economically viable way to step down the voltage of DC for end user applications such as lighting incandescent bulbs.

Three-phase electrical generation is very common. The simplest way is to use three separate coils in the generator stator, physically offset by an angle of 120° (one-third of a complete 360° phase) to each other. Three current waveforms are produced that are equal in magnitude and 120° out of phase to each other. If coils are added opposite to these (60° spacing), they generate the same phases with reverse polarity and so can be simply wired together. In practice, higher "pole orders" are commonly used. For example, a 12-pole machine would have 36 coils (10° spacing). The advantage is that lower rotational speeds can be used to generate the same frequency. For example, a 2-pole machine running at 3600 rpm and a 12-pole machine running at 600 rpm produce the same frequency; the lower speed is preferable for larger machines. If the load on a three-phase system is balanced equally among the phases, no current flows through the neutral point. Even in the worst-case unbalanced (linear) load, the neutral current will not exceed the highest of the phase currents. Non-linear loads (e.g., the switch-mode power supplies widely used) may require an oversized neutral bus and neutral conductor in the upstream distribution panel to handle harmonics. Harmonics can cause neutral conductor current levels to exceed that of one or all phase conductors.

For three-phase at utilization voltages a four-wire system is often used. When stepping down three-phase, a transformer with a Delta (3-wire) primary and a Star (4-wire, center-earthed) secondary is often used so there is no need for a neutral on the supply side. For smaller customers (just how small varies by country and age of the installation) only a single phase and neutral, or two phases and neutral, are taken to the property. For larger installations all three phases and neutral are taken to the main distribution panel. From the three-phase main panel, both single and three-phase circuits may lead off. Three-wire single-phase systems, with a single center-tapped transformer giving two live conductors, is a common distribution scheme for residential and small commercial buildings in North America. This arrangement is sometimes incorrectly referred to as "two phase". A similar method is used for a different reason on construction sites in the UK. Small power tools and lighting are supposed to be supplied by a local center-tapped transformer with a voltage of 55 V between each power conductor and earth. This significantly reduces the risk of electric shock in the event that one of the live conductors becomes exposed through an equipment fault whilst still allowing a reasonable voltage of 110 V between the two conductors for running the tools.

A third wire, called the bond (or earth) wire, is often connected between non-current-carrying metal enclosures and earth ground. This conductor provides protection from electric shock due to accidental contact of circuit conductors with the metal chassis of portable appliances and tools. Bonding all non-current-carrying metal parts into one complete system ensures there is always a low electrical impedance path to ground sufficient to carry any fault current for as long as it takes for the system to clear the fault. This low impedance path allows the maximum amount of fault current, causing the overcurrent protection device (breakers, fuses) to trip or burn out as quickly as possible, bringing the electrical system to a safe state. All bond wires are bonded to ground at the main service panel, as is the Neutral/Identified conductor if present.

AC Power Supply Frequencies

The frequency of the electrical system varies by country and sometimes within a country; most electric power is generated at either 50 or 60 hertz. Some countries have a mixture of 50 Hz and 60 Hz supplies, notably electricity power transmission in Japan. A low frequency eases the design of electric motors, particularly for hoisting, crushing and rolling applications, and commutator-type traction motors for applications such as railways. However, low frequency also causes noticeable flicker in arc lamps and incandescent light bulbs. The use of lower frequencies also provided the advantage of lower impedance losses, which are proportional to frequency. The original Niagara Falls generators were built to produce 25 Hz power, as a compromise between low frequency for traction and heavy induction motors, while still allowing incandescent lighting to operate (although with noticeable flicker). Most of the 25 Hz residential and commercial customers for Niagara Falls power were converted to 60 Hz by the late 1950s, although some 25 Hz industrial customers still existed as of the start of the 21st century. 16.7 Hz power (formerly 16 2/3 Hz) is still used in some European rail systems, such as in Austria, Germany, Norway, Sweden and Switzerland. Off-shore, military, textile industry, marine, aircraft, and spacecraft applications sometimes use 400 Hz, for benefits of reduced weight of apparatus or higher motor speeds. Computer mainframe systems are often powered by 415 Hz, using customer-supplied 35 or 70 KVA motor-generator sets. Smaller mainframes may have an internal 415 Hz M-G set. In any case, the input to the M-G set is the local customary voltage and frequency, variously 200 (Japan), 208, 240 (North America), 380, 400 or 415 (Europe) volts, and variously 50 or 60 Hz.

Effects at High Frequencies

A Tesla coil producing high-frequency current that is harmless to humans, but lights a neon tube when brought near it. Performed by Prof. Oliver Zajkov at the Physics Institute at the Ss. Cyril and Methodius University of Skopje, Macedonia.

A direct current flows uniformly throughout the cross-section of a uniform wire. An alternating current of any frequency is forced away from the wire's center, toward its outer surface. This is because the acceleration of an electric charge in an alternating current produces waves of electromagnetic radiation that cancel the propagation of electricity toward the center of materials with high conductivity. This phenomenon is called skin effect. At very high frequencies the current no longer flows *in* the wire, but effectively flows *on* the surface of the wire, within a thickness of a few skin depths. The skin depth is the thickness at which the current density is reduced by 63%. Even at relatively low frequencies used for power transmission (50–60 Hz), non-uniform distribution of current still occurs in sufficiently thick conductors. For example, the skin depth of a copper conductor is approximately 8.57 mm at 60 Hz, so high current conductors are usually hollow to reduce their mass and cost. Since the current tends to flow in the periphery of conductors, the effective cross-section of the conductor is reduced. This increases the effective AC resistance of the conductor, since resistance is inversely proportional to the cross-sectional area. The AC resistance often is many times higher than the DC resistance, causing a much higher energy loss due to ohmic heating (also called I^2R loss).

Techniques for Reducing AC Resistance

For low to medium frequencies, conductors can be divided into stranded wires, each insulated from one other, and the relative positions of individual strands specially arranged within the conductor bundle. Wire constructed using this technique is called Litz wire. This measure helps to partially mitigate skin effect by forcing more equal current throughout the total cross section of the stranded conductors. Litz wire is used for making high-Q inductors, reducing losses in flexible conductors carrying very high currents at lower frequencies, and in the windings of devices carrying higher radio frequency current (up to hundreds of kilohertz), such as switch-mode power supplies and radio frequency transformers.

Techniques for Reducing Radiation Loss

As written above, an alternating current is made of electric charge under periodic acceleration, which causes radiation of electromagnetic waves. Energy that is radiated is lost. Depending on the frequency, different techniques are used to minimize the loss due to radiation.

Twisted pairs

At frequencies up to about 1 GHz, pairs of wires are twisted together in a cable, forming a twisted pair. This reduces losses from electromagnetic radiation and inductive coupling. A twisted pair must be used with a balanced signalling system, so that the two wires carry equal but opposite currents. Each wire in a twisted pair radiates a signal, but it is effectively cancelled by radiation from the other wire, resulting in almost no radiation loss.

Coaxial Cables

Coaxial cables are commonly used at audio frequencies and above for convenience. A coaxial cable has a conductive wire inside a conductive tube, separated by a dielectric layer. The current flowing on the inner conductor is equal and opposite to the current flowing on the inner surface of the tube. The electromagnetic field is thus completely contained within the tube, and (ideally) no energy is lost to radiation or coupling outside the tube. Coaxial cables have acceptably small losses for frequencies up to about 5 GHz. For microwave frequencies greater than 5 GHz, the losses (due mainly to the electrical resistance of the central conductor) become too large, making waveguides a more efficient medium for transmitting energy. Coaxial cables with an air rather than solid dielectric are preferred as they transmit power with lower loss.

Waveguides

Waveguides are similar to coax cables, as both consist of tubes, with the biggest difference being that the waveguide has no inner conductor. Waveguides can have any arbitrary cross section, but rectangular cross sections are the most common. Because waveguides do not have an inner conductor to carry a return current, waveguides cannot deliver energy by means of an electric current, but rather by means of a *guided* electromagnetic field. Although surface currents do flow on the inner walls of the waveguides, those surface currents do not carry power. Power is carried by the guided electromagnetic fields. The surface currents are set up by the guided electromagnetic fields and have the effect of keeping the fields inside the waveguide and preventing leakage of the fields to the space outside the waveguide. Waveguides have dimensions comparable to the wavelength of the alternating current to be transmitted, so they are only feasible at microwave frequencies. In addition to this mechanical feasibility, electrical resistance of the non-ideal metals forming the walls of the waveguide cause dissipation of power (surface currents flowing on lossy conductors dissipate power). At higher frequencies, the power lost to this dissipation becomes unacceptably large.

Fibre Optics

At frequencies greater than 200 GHz, waveguide dimensions become impractically small, and the ohmic losses in the waveguide walls become large. Instead, fibre optics, which are a form of dielectric waveguides, can be used. For such frequencies, the concepts of voltages and currents are no longer used.

Mathematics of AC Voltages

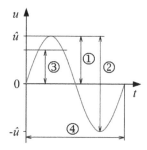

A sinusoidal alternating voltage.
1 = peak, also amplitude,
2 = peak-to-peak,
3 = effective value,
4 = Period

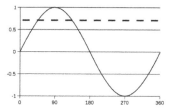

A sine wave, over one cycle (360°). The dashed line represents the root mean square (RMS) value at about 0.707

Alternating currents are accompanied (or caused) by alternating voltages. An AC voltage v can be described mathematically as a function of time by the following equation:

$$v(t) = V_{peak} \cdot \sin(\omega t),$$

where

- V_{peak} is the peak voltage (unit: volt),

- ω is the angular frequency (unit: radians per second)

  - The angular frequency is related to the physical frequency, f (unit = hertz), which represents the number of cycles per second, by the equation $\omega = 2\pi f$.

- t is the time (unit: second).

The peak-to-peak value of an AC voltage is defined as the difference between its positive peak and its negative peak. Since the maximum value of $\sin(x)$ is +1 and the minimum value is −1, an AC voltage swings between $+V_{peak}$ and $-V_{peak}$. The peak-to-peak voltage, usually written as V_{pp} or V_{p-p}, is therefore $V_{peak} - (-V_{peak}) = 2V_{peak}..$

Power

The relationship between voltage and the power delivered is

$p(t) = \dfrac{v^2(t)}{R}$ where R represents a load resistance.

Rather than using instantaneous power, $p(t)$, it is more practical to use a time averaged power (where the averaging is performed over any integer number of cycles). Therefore, AC voltage is often expressed as a root mean square (RMS) value, written as V_{rms},, because

$$P_{time\ averaged} = \frac{V^2_{rms}}{R}.$$

Power oscillation

$$v(t) = V_{peak} \sin(\omega t)$$

$$i(t) = \frac{v(t)}{R} = \frac{V_{peak}}{R} \sin(\omega t)$$

$$P(t) = v(t)\, i(t) = \frac{(V_{peak})^2}{R} \sin^2(\omega t)$$

Information Transmission

Alternating current is used to transmit information, as in the cases of telephone and cable television. Information signals are carried over a wide range of AC frequencies. POTS telephone signals have a frequency of about 3 kilohertz, close to the baseband audio frequency. Cable television and other cable-transmitted information currents may alternate at frequencies of tens to thousands of megahertz. These frequencies are similar to the electromagnetic wave frequencies often used to transmit the same types of information over the air.

History

The first alternator to produce alternating current was a dynamo electric generator based on Michael Faraday's principles constructed by the French instrument maker Hippolyte Pixii in 1832. Pixii later added a commutator to his device to produce the (then) more commonly used direct current. The earliest recorded practical application of alternating current is by Guillaume Duchenne, inventor and developer of electrotherapy. In 1855, he announced that AC was superior to direct current for electrotherapeutic triggering of muscle contractions. Alternating current technology had first developed in Europe due to the work of Guillaume Duchenne (1850s), the Hungarian Ganz Works company (1870s), and in the 1880s: Sebastian Ziani de Ferranti, Lucien Gaulard, and Galileo Ferraris.

In 1876, Russian engineer Pavel Yablochkov invented a lighting system based on a set of induction coils where the primary windings were connected to a source of AC. The secondary windings could be connected to several 'electric candles' (arc lamps) of his own design. The coils Yablochkov employed functioned essentially as transformers. In 1878, the Ganz factory, Budapest, Hungary, began manufacturing equipment for electric lighting and, by 1883, had installed over fifty systems in Austria-Hungary. Their AC systems used arc and incandescent lamps, generators, and other equipment. A power transformer developed by Lucien Gaulard and John Dixon Gibbs was demonstrated in London in 1881, and attracted the interest of Westinghouse. They also exhibited the invention in Turin in 1884.

DC Distribution Systems

During the initial years of electricity distribution, Edison's direct current was the standard for the United States, and Edison did not want to lose all his patent royalties. Direct current worked well with incandescent lamps, which were the principal electric load of the day, and with electric motors. Direct-current systems could be directly used with storage batteries, providing valuable load-leveling and backup power during interruptions of generator operation. Direct-current generators could be easily paralleled, allowing economical operation by using smaller machines during periods of light load and improving reliability. At the introduction of Edison's system, no practical AC motor was available. Edison had invented a meter to allow customers to be billed for energy proportional to consumption, but this meter worked only with direct current. The principal drawback of direct-current distribution was that customer loads, distribution and generation were all at the same voltage. Generally, it was uneconomical to use a high voltage for transmission and reduce it for customer uses. Even with the Edison 3-wire system (placing two 110-volt customer loads in series on a 220-volt supply), the high cost of conductors required generation to be close to customer loads, otherwise losses made the system uneconomical to operate.

Transformers

Alternating current systems can use transformers to change voltage from low to high level and back, allowing generation and consumption at low voltages but transmission, possibly over great distances, at high voltage, with savings in the cost of conductors and energy losses. A bipolar open-core power transformer developed by Lucien Gaulard and John Dixon Gibbs was demonstrated in London in 1881, and attracted the interest of Westinghouse. They also exhibited the invention in Turin in 1884. However these early induction coils with open magnetic circuits are inefficient at transferring power to loads. Until about 1880, the paradigm for AC power transmission from a high voltage supply to a low voltage load was a series circuit. Open-core transformers with a ratio near 1:1 were connected with their primaries in series to allow use of a high voltage for transmission while presenting a low voltage to the lamps. The inherent flaw in this method was that turning off a single lamp (or other electric device) affected the voltage supplied to all others on the same circuit. Many adjustable transformer designs were introduced to compensate for this problematic characteristic of the series circuit, including those employing methods of adjusting the core or bypassing the magnetic flux around part of a coil. The direct current systems did not have these drawbacks, giving it significant advantages over early AC systems.

Pioneers

In the autumn of 1884, Károly Zipernowsky, Ottó Bláthy and Miksa Déri (ZBD), three engineers associated with the Ganz factory, determined that open-core devices were impractical, as they were incapable of reliably regulating voltage. In their joint 1885 patent applications for novel transformers (later called ZBD transformers), they described two designs with closed magnetic circuits where copper windings were either a) wound around iron wire ring core or b) surrounded by iron wire core. In both designs, the magnetic flux linking the primary and secondary windings traveled almost entirely within the confines of the iron core, with no intentional path through air. The new transformers were 3.4 times more efficient than the open-core bipolar devices of Gaulard and Gibbs. The Ganz factory in 1884 shipped the world's first five high-efficiency AC transformers. This first unit had been manufactured to the following specifications: 1,400 W, 40 Hz, 120:72 V, 11.6:19.4 A, ratio 1.67:1, one-phase, shell form.

The prototype of ZBD. transformer is on display at the Széchenyi István Memorial Exhibition, Nagycenk, Hungary

The ZBD patents included two other major interrelated innovations: one concerning the use of parallel connected, instead of series connected, utilization loads, the other concerning the ability to have high turns ratio transformers such that the supply network voltage could be much higher (initially 1,400 to 2,000 V) than the voltage of utilization loads (100 V initially preferred). When employed in parallel connected electric distribution systems, closed-core transformers finally made it technically and economically feasible to provide electric power for lighting in homes, businesses and public spaces. The other essential milestone was the introduction of 'voltage source, voltage intensive' (VSVI) systems' by the invention of constant voltage generators in 1885. Ottó Bláthy also invented the first AC electricity meter.

The Hungarian "ZBD" Team (Károly Zipernowsky, Ottó Bláthy, Miksa Déri). They were the inventors of the first high efficiency, closed core shunt connection transformer. The three also invented the modern power distribution system: Instead of former series connection they connect transformers that supply the appliances in parallel to the main line. Blathy invented the AC Wattmeter, and they invented the essential Constant Voltage Generator.

The AC power systems was developed and adopted rapidly after 1886 due to its ability to distribute electricity efficiently over long distances, overcoming the limitations of the direct current system. In 1886, the ZBD engineers designed, and the Ganz factory supplied electrical equipment for, the world's first power station that used AC generators to power a parallel connected common electrical network, the steam-powered Rome-Cerchi power plant. The reliability of the AC technology received impetus after the Ganz Works electrified a large European metropolis: Rome in 1886.

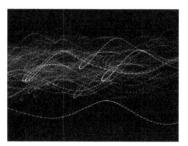

The city lights of Prince George, British Columbia viewed in a motion blurred exposure. The AC blinking causes the lines to be dotted rather than continuous.

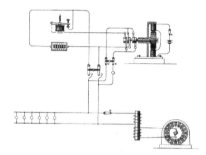

Westinghouse Early AC System 1887
(US patent 373035)

In the UK, Sebastian de Ferranti, who had been developing AC generators and transformers in London since 1882, redesigned the AC system at the Grosvenor Gallery power station in 1886 for the London Electric Supply Corporation (LESCo) including alternators of his own design and transformer designs similar to Gaulard and Gibbs. In 1890 he designed their power station at Deptford and converted the Grosvenor Gallery station across the Thames into an electrical substation, showing the way to integrate older plants into a universal AC supply system.

In the US William Stanley, Jr. designed one of the first practical devices to transfer AC power efficiently between isolated circuits. Using pairs of coils wound on a common iron core, his design, called an induction coil, was an early (1885) transformer. Stanley also worked on engineering and adapting European designs such as the Gaulard and Gibbs transformer for US entrepreneur George Westinghouse who started building AC systems in 1886. The spread of Westinghouse and other AC systems triggered a push back in late 1887 by Thomas Edison (a proponent of direct current) who attempted to discredit alternating current as too dangerous in a public campaign called the "War of Currents". In 1888 alternating current systems gained further viability with introduction of a functional AC motor, something these systems had lacked up till then. The design, an induction motor, was independently invented by Galileo Ferraris and Nikola Tesla (with Tesla's design being licensed by Westinghouse in the US). This design was further developed into the modern practical three-phase form by Mikhail Dolivo-Dobrovolsky and Charles Eugene Lancelot Brown.

The Ames Hydroelectric Generating Plant (spring of 1891) and the original Niagara Falls Adams Power Plant (August 25, 1895) were among the first hydroelectric AC-power plants. The first commercial power plant in the United States using three-phase alternating current was the hydroelectric Mill Creek No. 1 Hydroelectric Plant near Redlands, California, in 1893 designed by Almirian Decker. Decker's design incorporated 10,000-volt three-phase transmission and established the standards for the complete system of generation, transmission and motors used today. The Jaruga Hydroelectric Power Plant in Croatia was set in operation on 28 August 1895. The two generators (42 Hz, 550 kW each) and the transformers were produced and installed by the Hungarian company Ganz. The transmission line from the power plant to the City of Šibenik was 11.5 kilometers (7.1 mi) long on wooden towers, and the municipal distribution grid 3000 V/110 V included six transforming stations. Alternating current circuit theory developed rapidly in the latter part of the 19th and early 20th century. Notable contributors to the theoretical basis of alternating current calculations include Charles Steinmetz, Oliver Heaviside, and many others. Calculations in unbalanced three-phase systems were simplified by the symmetrical components methods discussed by Charles Legeyt Fortescue in 1918.

Electric Machine

In electrical engineering, electric machine is a general term for electric motors and electric generators. They are electromechanical energy converters: an electric motor converts electricity to mechanical power while an electric generator converts mechanical power to electricity. The moving parts in a machine can be rotating (*rotating machines*) or linear (*linear machines*). Besides motors and generators, a third category often included is transformers, which although they do not have any moving parts are also energy converters, changing the voltage level of an alternating current.

Electric machines, in the form of generators, produce virtually all electric power on Earth, and in the form of electric motors consume approximately 60% of all electric power produced. Electric machines were developed beginning in the mid 19th century and since that time have been a ubiquitous component of the infrastructure. Developing more efficient electric machine technology is crucial to any global conservation, green energy, or alternative energy strategy.

Classifications

When classifying electric machines (motors and generators) it is reasonable to start with physical principle for converting electric energy to mechanical energy. If the controller is included as a part of the machine all machines can be powered by either alternating or direct current, although some machines will need a more advanced controller than others. Classification is complicated by the possibilities of combining physical principles when constructing an electrical machine. It can, for example, be possible to run a brushed machine as a reluctance machine (without using the rotor coils) if the rotor iron has the correct shape.

Generally all electric machines can be turned inside out, so rotor and stator exchange places. All rotating electric machines have an equivalent linear electric machine where stator moves along a straight line instead of rotating. The opposite—linear to rotary dual—is not always the case. Motors and generators can be designed with or without iron to improve the path of the magnetic field (teeth to reduce the air gap is a common example) and with and without permanent magnets (PM), with different pole number etc., but still belong to different classes of machines. Electric machines can be synchronous meaning that the magnetic field set up by the stator coils rotates with the same speed as the rotor; or asynchronous, meaning that there is a speed difference. PM machines and reluctance machines are always synchronous. Brushed machines with rotor windings can be synchronous when the rotor is supplied with DC or AC with same frequency as stator or asynchronous when stator and rotor are supplied with AC with different frequencies. Induction machines are usually asynchronous, but can be synchronous, if there are superconductors in the rotor windings.

Comparing Electric Machine Systems

Comparing the cost-performance between electric machine systems of different classification or from different manufacturers is difficult without a critical baseline of metrics. For equitable comparison of efficiency, cost, torque, and power between electric machine systems, the comparison should be matched with the same voltage and speed at a given frequency of excitation; or instead, the additional cost, efficiency, and real-estate of the transmission for instance for coupling to the speed of the application, the transformer for matching voltage, the frequency converter to match excitation frequency, etc., should be included.

Other parameters that should always be considered in any revealing comparison:

- Duty Cycle – although directly related to cost, efficiency, and power density, duty cycle gives meaning to application applicability;

- Peak Torque Potential – the closeness of peak torque potential to continuous torque indicates the safe margin of the design;

- Cost, Efficiency, and Real-estate of the Electronic Controller – unless integrated and included in the specifications of the electric machine system, the electronic controller, which is required for practical system operation, should always be included.

- Utilization of the Magnetic Core and frame assembly – As an example: with the wound-rotor doubly fed electric machine as the only exception, rotor assemblies, which consume nearly half the volume of the electric machine, *passively* participate in the energy conversion process and are under-utilized.

Generator

Electric generator.

An electric generator is a device that converts mechanical energy to electrical energy. A generator forces electrons to flow through an external electrical circuit. It is somewhat analogous to a water pump, which creates a flow of water but does not create the water inside. The source of mechanical energy, the prime mover, may be a reciprocating or turbine steam engine, water falling through a turbine or waterwheel, an internal combustion engine, a wind turbine, a hand crank, compressed air or any other source of mechanical energy.

The two main parts of an electrical machine can be described in either mechanical or electrical terms. In mechanical terms, the rotor is the rotating part, and the stator is the stationary part of an electrical machine. In electrical terms, the armature is the power-producing component and the field is the magnetic field component of an electrical machine. The armature can be on either the rotor or the stator. The magnetic field can be provided by either electromagnets or permanent magnets mounted on either the rotor or the stator. Generators are classified into two types, AC generators and DC generators.

AC Generator

An AC generator converts mechanical energy into alternating current electricity. Because power transferred into the field circuit is much less than power transferred into the armature circuit, AC generators nearly always have the field winding on the rotor and the armature winding on the stator.

AC generators are classified into several types.

In an induction generator, the stator magnetic flux induces currents in the rotor. The prime mover then drives the rotor above the synchronous speed, causing the opposing rotor flux to cut the stator coils producing active current in the stator coils, thus sending power back to the electrical grid. An induction generator draws reactive power from the connected system and so cannot be an isolated source of power.

In a Synchronous generator (alternator), the current for the magnetic field is provided by a separate DC current source.

DC Generator

An electric generator is a machine that converts mechanical energy into electrical energy. An electric generator works based on the principle that whenever conductor cuts the magnetic field, an emf is induced which will cause the current to flow if conductor circuit is closed. Take the working of Simple loop Generator as an example.

Motor

An electric motor converts electrical energy into mechanical energy. The reverse process of electrical generators, most electric motors operate through interacting magnetic fields and current-carrying conductors to generate rotational force. Motors and generators have many similarities and many types of electric motors can be run as generators, and vice versa. Electric motors are found in applications as diverse as industrial fans, blowers and pumps, machine tools, household appliances, power tools, and disk drives. They may be powered by direct current or by alternating current which leads to the two main classifications: AC motors and DC motors.

Electric motor.

AC Motor

An AC motor converts alternating current into mechanical energy. It commonly consists of two basic parts, an outside stationary stator having coils supplied with alternating current to produce a rotating magnetic field, and an inside rotor attached to the output shaft that is given a torque by the rotating field. The two main types of AC motors are distinguished by the type of rotor used.

- Induction (asynchronous) motor, the rotor magnetic field is created by an induced current. The rotor must turn slightly slower (or faster) than the stator magnetic field to provide the induced current. There are three types of induction motor rotors, which are squirrel-cage rotor, wound rotor and solid core rotor.

- Synchronous motor, it does not rely on induction and so can rotate exactly at the supply frequency or sub-multiple. The magnetic field of the rotor is either generated by direct current delivered through slip rings (exciter) or by a permanent magnet.

DC Motor

The brushed DC electric motor generates torque directly from DC power supplied to the motor by using internal commutation, stationary permanent magnets, and rotating electrical magnets. Brushes and springs carry the electric current from the commutator to the spinning wire windings of the rotor inside the motor. Brushless DC motors use a rotating permanent magnet in the rotor, and stationary electrical magnets on the motor housing. A motor controller converts DC to AC. This design is simpler than that of brushed motors because it eliminates the complication of transferring power from outside the motor to the spinning rotor. An example of a brushless, synchronous DC motor is a stepper motor which can divide a full rotation into a large number of steps.

Transformer

A transformer is a static device that converts alternating current from one voltage level to another level (higher or lower), or to the same level, without changing the frequency. A transformer transfers electrical energy from one circuit to another through inductively coupled conductors—the transformer's coils. A varying electric current in the first or *primary* winding creates a varying magnetic flux in the transformer's core and thus a varying magnetic field through the *secondary* winding. This varying magnetic field induces a varying electromotive force (EMF) or "voltage" in the secondary winding. This effect is called mutual induction.

Transformer.

There are three types of transformers

1. Step-up transformer

2. Step-down transformer

3. Isolation transformer

There are four types of transformers based on structure

1. core type

2. shell type

3. power type

4. instrument type

Electromagnetic-rotor Machines

Electromagnetic-rotor machines are machines having some kind of electric current in the rotor which creates a magnetic field which interacts with the stator windings. The rotor current can be the internal current in a permanent magnet (PM machine), a current supplied to the rotor through brushes (Brushed machine) or a current set up in closed rotor windings by a varying magnetic field (Induction machine).

Permanent Magnet Machines

PM machines have permanent magnets in the rotor which set up a magnetic field. The magnetomotive force in a PM (caused by orbiting electrons with aligned spin) is generally much higher than what is possible in a copper coil. The copper coil can, however, be filled with a ferromagnetic material, which gives the coil much lower magnetic reluctance. Still the magnetic field created by modern PMs (Neodymium magnets) is stronger, which means that PM machines have a better torque/volume and torque/weight ratio than machines with rotor coils under continuous operation. This may change with introduction of superconductors in rotor.

Since the permanent magnets in a PM machine already introduce considerable magnetic reluctance, then the reluctance in the air gap and coils are less important. This gives considerable freedom when designing PM machines.

It is usually possible to overload electric machines for a short time until the current in the coils heats parts of the machine to a temperature which cause damage. PM machines can in less degree be subjected to such overload because too high current in the coils can create a magnetic field strong enough to demagnetise the magnets.

Brushed Machines

Brushed machines are machines where the rotor coil is supplied with current through brushes in much the same way as current is supplied to the car in an electric slot car track. More durable brushes can be made of graphite or liquid metal. It is even possible to eliminate the brushes in a "brushed machine" by using a part of rotor and stator as a transformer which transfer current without creating torque. Brushes must not be confused with a commutator. The difference is that the brushes only transfer electric current to a moving rotor while a commutator also provide switching of the current direction.

There is iron (usually laminated steel cores made of sheet metal) between the rotor coils and teeth of iron between the stator coils in addition to black iron behind the stator coils. The gap between rotor and stator is also made as small as possible. All this is done to minimize magnetic reluctance of the magnetic circuit which the magnetic field created by the rotor coils travels through, something which is important for optimizing these machines.

Large brushed machines which are run with DC to the stator windings at synchronous speed are the most common generator in power plants, because they also supply reactive power to the grid, because they can be started by the turbine and because the machine in this system can generate power at constant speed without a controller. This type of machine is often referred to in the literature as a synchronous machine.

This machine can also be run by connecting the stator coils to the grid, and supplying the rotor coils with AC from an inverter. The advantage is that it is possible to control rotating speed of the machine with a fractionally rated inverter. When run this way the machine is known as a brushed double feed "induction" machine. "Induction" is misleading because there is no useful current in the machine which is set up by induction.

Induction Machines

Induction machines have short circuited rotor coils where a current is set up and maintained by induction. This requires that the rotor rotates at other than synchronous speed, so that the rotor coils are subjected to a varying magnetic field created by the stator coils. An induction machine is an asynchronous machine.

Induction eliminates the need for brushes which is usually a weak part in an electric machine. It also allows designs which make it very easy to manufacture the rotor. A metal cylinder will work as rotor, but to improve efficiency a "squirrel cage" rotor or a rotor with closed windings is usually used. The speed of asynchronous induction machines will decrease with increased load because a larger speed difference between stator and rotor is necessary to set up sufficient rotor current and rotor magnetic field. Asynchronous induction machines can be made so they start and run without any means of control if connected to an AC grid, but the starting torque is low.

A special case would be an induction machine with superconductors in the rotor. The current in the superconductors will be set up by induction, but the rotor will run at synchronous speed because there will be no need for a speed difference between the magnetic field in stator and speed of rotor to maintain the rotor current.

Another special case would be the brushless double fed induction machine, which has a double set of coils in the stator. Since it has two moving magnetic fields in the stator, it gives no meaning to talk about synchronous or asynchronous speed.

Reluctance Machines

Reluctance machines have no windings in rotor, only a ferromagnetic material shaped so that "electromagnets" in stator can "grab" the teeth in rotor and move it a little. The electromagnets are then turned off, while another set of electromagnets is turned on to move stator further. Another name is step motor, and it is suited for low speed and accurate position control. Reluctance machines can be supplied with PMs in stator to improve performance. The "electromagnet" is then "turned off" by sending a negative current in the coil. When the current is positive the magnet and the current cooperate to create a stronger magnetic field which will improve the reluctance machine's maximum torque without increasing the currents maximum absolute value.

Electrostatic Machines

In *electrostatic machines*, torque is created by attraction or repulsion of electric charge in rotor and stator.

Electrostatic generators generate electricity by building up electric charge. Early types were friction machines, later ones were influence machines that worked by electrostatic induction. The Van de Graaff generator is an electrostatic generator still used in research today.

Homopolar Machines

Homopolar machines are true DC machines where current is supplied to a spinning wheel through brushes. The wheel is inserted in a magnetic field, and torque is created as the current travels from the edge to the centre of the wheel through the magnetic field.

Electric Machine Systems

For optimized or practical operation of electric machines, today's electric machine systems are complemented with electronic control.

Passive Sign Convention

In electrical engineering, the passive sign convention (PSC) is a sign convention or arbitrary standard rule adopted universally by the electrical engineering community for defining the sign of electric power in an electric circuit. The convention defines electric power flowing out of the circuit *into* an electrical component as positive, and power flowing into the circuit *out* of a component as negative. So a passive component which consumes power, such as an appliance or light bulb, will have *positive* power dissipation, while an active component, a source of power such as an electric generator or battery, will have *negative* power dissipation. This is the standard definition of power in electric circuits; it is used for example in computer circuit simulation programs such as SPICE.

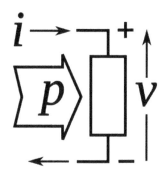

Illustration of the "*reference directions*" of the current (*i*), voltage (*v*), and power (*p*) variables used in the passive sign convention. If positive current is defined as flowing into the terminal which is defined to have positive voltage, then positive power represents electric power flowing into the device (*big arrow*).

To comply with the convention, the direction of the voltage and current variables used to calculate power and resistance in the component must have a certain relationship: the current variable must be defined so positive current enters the positive voltage terminal of the device. These directions may be different from the directions of the actual current flow and voltage.

The Convention

The passive sign convention states that in components in which the conventional current variable *i* is defined as entering the device through the terminal which is positive as defined by the voltage variable *v*, the power *p* and resistance *r* are given by

$$p = vi \quad \text{and} \quad r = v/i$$

In components in which the current *i* enters the device through the negative voltage terminal, power and resistance are given by

$$p = -vi \quad \text{and} \quad r = -v/i$$

With these definitions, passive components (loads) will have $p > 0$ and $r > 0$, and active components (power sources) will have $p < 0$ and $r < 0$.

Explanation

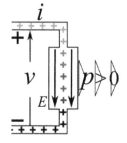

Load (passive component)

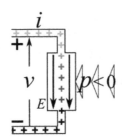

Power source (active component)

Active and Passive Components

In electrical engineering, power represents the rate of electrical energy flowing into or out of a given component or control volume. Power is a signed quantity; negative power just represents power flowing in the opposite direction from positive power. From the standpoint of power flow, electrical components in a circuit can be divided into two types:

- In a *load* or *passive* component, such as a light bulb, resistor, or electric motor, electric current (flow of positive charges) moves through the device under the influence of the voltage in the direction of lower electric potential, from the positive terminal to the negative. So work is done *by* the charges *on* the component; potential energy flows out of the charges; and electric power flows from the circuit into the component, where it is converted to some other form of energy such as heat or mechanical work.

- In a *source* or *active* component, such as a battery or electric generator, current is forced to move through the device in the direction of greater electric potential energy, from the negative to the positive voltage terminal. This increases the potential energy of the electric charges, so electric power flows out of the component into the circuit. Work must be done *on* the moving charges by some source of energy in the component, to make them move in this direction against the opposing force of the electric field E.

Some components can be either a source or a load, depending on the voltage or current through them. For example a rechargeable battery acts as a source when it is used to produce power, but as a load when it is being recharged.

Since it can flow in either direction, there are two possible ways to define electric power; two possible *reference directions*: either power flowing into the circuit, or power flowing out of the circuit, can be defined as positive. Whichever is defined as positive, the other will be negative. The passive sign convention arbitrarily defines power flowing *out* of the circuit (*into* the component) as positive, so passive components have "positive" power flow.

Reference Directions

The power flow p and resistance r of an electrical component are related to the voltage v and current i variables by the defining equation for power and Ohm's law:

$$p = vi \qquad (1)$$

$$r = v / i \qquad\qquad (2)$$

Like power, voltage and current are signed quantities. The current flow in a wire has two possible directions, so when defining a current variable i the direction which represents positive current flow must be indicated, usually by an arrow on the circuit diagram. This is called the *reference direction for current i*. If the actual current is in the opposite direction, the variable i will have a negative value. Similarly in defining a voltage variable v, the terminal which represents the positive side must be specified, usually with an arrow or plus sign. This is called the *reference direction for voltage v*.

To understand the PSC, it is important to distinguish the reference directions of the variables, v and i, which can be assigned at will, from the direction of the actual *voltage* and *current*, which is determined by the circuit. The idea of the PSC is that by assigning the reference direction of variables v and i in a component with the right relationship, the power flow in passive components calculated from Eq. (1) will come out positive, while the power flow in active components will come out negative. It is not necessary to know whether a component produces or consumes power when analyzing the circuit; reference directions can be assigned arbitrarily, directions to currents and polarities to voltages, then the PSC is used to calculate the power in components. If the power comes out positive, the component is a load, converting electric power to some other kind of power. If the power comes out negative, the component is a source, converting some other form of power to electric power.

Sign Conventions

The above discussion shows that choosing the relative direction of the voltage and current variables in a component determines the direction of power flow that is considered positive. The reference directions of the individual variables are not important, only their relation to each other. There are two choices:

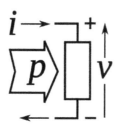

- *Passive sign convention*: Defining the current variable as entering the positive terminal means that if the voltage and current variables have positive values, current flows from the positive to the negative terminal, doing work *on* the component, as occurs in a passive component. So power flowing *into* the component from the line is defined as positive; the power variable represents power *dissipation* in the component. Therefore

    - Active components (power sources) will have negative resistance and negative power flow

    - Passive components (loads) will have positive resistance and positive power flow

This is the convention normally used.

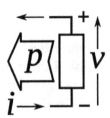

- *Active sign convention*: Defining the current variable as entering the negative terminal means that if the voltage and current variables have positive values, current flows from the negative to the positive terminal, so work is being done *on* the current, and power flows *out* of the component. So power flowing out of the component is defined as positive; the power variable represents power *produced*. Therefore:

  - Active components will have positive resistance and positive power flow
  - Passive components will have negative resistance and negative power flow

This convention is rarely used, except for special cases in power engineering.

In practice it is not necessary to assign the voltage and current variables in a circuit to comply with the PSC. Components in which the variables have a "backward" relationship, in which the current variable enters the negative terminal, can still be made to comply with the PSC by changing the sign of the constitutive relations (1) and (2) used with them. A current entering the negative terminal is equivalent to a negative current entering the positive terminal, so in such a component

$p = v(-i) = -vi$, and

$r = v / (-i) = -v / i$

Conservation of Energy

One advantage of defining all the variables in a circuit to comply with the PSC is that it makes it easy to express conservation of energy. Since electric energy cannot be created or destroyed, at any given instant every watt of power consumed by a load component must be produced by some source component in the circuit. Therefore the sum of all the power consumed by loads equals the sum of all the power produced by sources. Since with the PSC the power dissipation in sources is negative and power dissipation in loads is positive, the sum of all the power dissipation in all the components in a circuit is always zero

$$\sum_n P_n = \sum_n v_n i_n = 0$$

AC Circuits

Since the sign convention only deals with the directions of the *variables* and not with the direction of the actual *current*, it also applies to alternating current (AC) circuits, in which the direction of the voltage and current periodically reverses. In AC circuits, even though the voltage and current reverse direction, a "formal" direction of current flow and voltage polarity are defined by considering the voltage and current direction in the first half of the cycle "positive". During the second half of the AC cycle, in a resistive AC circuit, both the voltage and current in the device reverse direction, so the sign of the voltage and current reverse. Since the power is the product of voltage and current, the two sign reversals cancel each other, and the sign of the power flow is unchanged.

Alternative Convention in Power Engineering

In practice, the power output of power sources such as batteries and generators is not given in negative numbers, as required by the passive sign convention. No manufacturer sells a "−5 kilowatt generator". The standard practice in electric power circuits is to use positive values for the power and resistance of power sources, as well as loads. This avoids confusion over the meaning of "negative power", and particularly "negative resistance". In order to make the power for both sources and loads come out positive, instead of the PSC, separate sign conventions must be used for sources and loads. These are called the "*generator-load conventions*"

Generator convention - In source components like generators and batteries, the variables *V* and *I* are defined according to the *active sign convention* above; the current variable is defined as entering the negative terminal of the device.

Load convention - In loads, the variables are defined according to the normal passive sign convention; the current variable is defined as entering the positive terminal.

Using this convention, positive power flow in source components is power *produced*, while positive power flow in load components is power *consumed*. As with the PSC, if the variables in a given component do not conform to the applicable convention, the component can still be made to conform by using negative signs in the constitutive equations (1) and (2)

$$P = -VI \quad \text{and} \quad R = -V/I$$

This convention may seem preferable to the PSC, since the power *P* and resistance *R* always have positive values. However it cannot be used in electronics, because it is not possible to classify some electronic components unambiguously as "sources" or "loads". Some electronic components may act as sources of power with negative resistance in some portions of their operating range, and as absorbers of power with positive resistance in other portions, or even in different portions of the AC cycle. Whether the component acts as a source or load may depend on the current *i* or voltage *v* in it, which is not known until the circuit is analyzed. For example, if the voltage across a rechargeable battery's terminals is less than its open-circuit voltage, it will act as a source, while if the voltage is greater it will act as a load and recharge. So it is necessary for power and resistance variables to be able to take on both positive and negative values.

AC Power

Power in an electric circuit is the rate of flow of energy past a given point of the circuit. In alternating current circuits, energy storage elements such as inductors and capacitors may result in periodic reversals of the direction of energy flow. The portion of power that, averaged over a complete cycle of the AC waveform, results in net transfer of energy in one direction is known as active power (sometimes also called real power). The portion of power due to stored energy, which returns to the source in each cycle, is known as reactive power.

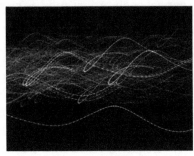

The blinking of non-incandescent city lights is shown in this motion-blurred long exposure. The AC nature of the mains power is revealed by the dashed appearance of the traces of moving lights.

Active, Reactive, and Apparent Power

In a simple alternating current (AC) circuit consisting of a source and a linear load, both the current and voltage are sinusoidal. If the load is purely resistive, the two quantities reverse their polarity at the same time. At every instant the product of voltage and current is positive or zero, with the result that the direction of energy flow does not reverse. In this case, only active power is transferred.

If the loads are purely *reactive*, then the voltage and current are 90 degrees out of phase. For half of each cycle, the product of voltage and current is positive, but on the other half of the cycle, the product is negative, indicating that on average, exactly as much energy flows toward the load as flows back. There is no net energy flow over one cycle. In this case, only reactive power flows— there is no net transfer of energy to the load.

Practical loads have resistance, inductance, and capacitance, so both active and reactive power will flow to real loads. Power engineers measure apparent power as the magnitude of the vector sum of active and reactive power. Apparent power is the product of the root-mean-square of voltage and current.

Electrical engineers take apparent power into account when designing and operating power systems, because though the current associated with reactive power does no work at the load, it heats the conductors and wastes energy. Conductors, transformers and generators must be sized to carry the total current, not just the current that does useful work. Failure to provide for the supply of sufficient reactive power in electrical grids can lead to lowered voltage levels and under certain operating conditions to the complete collapse of the network or blackout.

Another consequence is that adding the apparent power for two loads will not accurately give the total apparent power unless they have the same displacement between current and voltage (the same power factor).

Conventionally, capacitors are considered to generate reactive power and inductors to consume it. If a capacitor and an inductor are placed in parallel, then the currents flowing through the inductor and the capacitor tend to cancel rather than add. This is the fundamental mechanism for controlling the power factor in electric power transmission; capacitors (or inductors) are inserted in a circuit to partially compensate reactive power 'consumed' by the load. Purely capacitive circuits supply reactive power with the current waveform leading the voltage waveform by 90 degrees, while purely inductive circuits absorb reactive power with the current waveform lagging the voltage waveform by 90 degrees. The result of this is that capacitive and inductive circuit elements tend to cancel each other out.

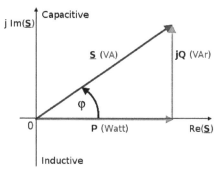

The complex power is the vector sum of active and reactive power. The apparent power is the magnitude of the complex power.
Active power, P
Reactive power, Q
Complex power, S
Apparent power, $|S|$
Phase of current, φ

Engineers use the following terms to describe energy flow in a system (and assign each of them a different unit to differentiate between them):

- Active power, P, or real power: watt (W)

- Reactive power, Q: volt-ampere reactive (var)

- Complex power, S: volt-ampere (VA)

- Apparent power, $|S|$: the magnitude of complex power S: volt-ampere (VA)

- Phase of voltage relative to current, φ: the angle of difference (in degrees) between current and voltage; current lagging voltage (quadrant I vector), current leading voltage (quadrant IV vector)

In the diagram, P is the active power, Q is the reactive power (in this case positive), S is the complex power and the length of S is the apparent power. Reactive power does not do any work, so it is represented as the imaginary axis of the vector diagram. Active power does do work, so it is the real axis.

The unit for all forms of power is the watt (symbol: W), but this unit is generally reserved for active power. Apparent power is conventionally expressed in volt-amperes (VA) since it is the product of rms voltage and rms current. The unit for reactive power is expressed as var, which stands for volt-ampere reactive. Since reactive power transfers no net energy to the load, it is sometimes called "wattless" power. It does, however, serve an important function in electrical grids and its lack has been cited as a significant factor in the Northeast Blackout of 2003.

Understanding the relationship among these three quantities lies at the heart of understanding power engineering. The mathematical relationship among them can be represented by vectors or expressed using complex numbers, $S = P + jQ$ (where j is the imaginary unit).

Power Factor

The ratio of active power to apparent power in a circuit is called the power factor. For two systems transmitting the same amount of active power, the system with the lower power factor will have higher circulating currents due to energy that returns to the source from energy storage in the load. These higher currents produce higher losses and reduce overall transmission efficiency. A lower power factor circuit will have a higher apparent power and higher losses for the same amount of active power.

The power factor is one when the voltage and current are in phase. It is zero when the current leads or lags the voltage by 90 degrees. Power factors are usually stated as "leading" or "lagging" to show the sign of the phase angle of current with respect to voltage. Voltage is designated as the base to which current angle is compared, meaning that we think of current as either "leading" or "lagging" voltage.

Where the waveforms are purely sinusoidal, the power factor is the cosine of the phase angle (φ) between the current and voltage sinusoid waveforms. Equipment data sheets and nameplates will often abbreviate power factor as " $\cos\phi$ " for this reason.

Example: The active power is 700 W and the phase angle between voltage and current is 45.6°. The power factor is $\cos(45.6°) = 0.700$. The apparent power is then: 700 W / $\cos(45.6°)$ = 1000 VA.

For instance, a power factor of 68 percent (0.68) means that only 68 percent of the total current supplied is actually doing work; the remaining 32 percent is reactive and has to be made up by the utility. Usually, utilities do not charge consumers for the reactive power losses as they do no real work for the consumer. However, if there are inefficiencies at the customer's load source that causes the power factor to fall below a certain level, utilities may charge customers in order to cover an increase in their power plant fuel use and their worse line and plant capacity.

Reactive Power

In a direct current circuit, the power flowing to the load is proportional to the product of the current through the load and the potential drop across the load. Energy flows in one direction from the source to the load.

In AC power, the voltage and current both vary approximately sinusoidally. When there is inductance or capacitance in the circuit, the voltage and current waveforms do not line up perfectly. The power flow has two components - one component flows from source to load and can perform work at the load, the other portion, known as "reactive power" is due to the delay between voltage and current, known as phase angle, and cannot do useful work at the load. It can be thought of as current that is arriving at the wrong time (too late or too early). To distinguish reactive power from active power, it is measured in units of "volt-amperes reactive" or var. These units can simplify to Watts, but are left as var to denote that they represent no actual work output.

Energy stored in capacitive or inductive elements of the network give rise to reactive power flow.

Reactive power flow strongly influences the voltage levels across the network. Voltage levels and reactive power flow must be carefully controlled to allow a power system to be operated within acceptable limits.

A technique known as reactive compensation is used to reduce apparent power flow to a load by reducing reactive power supplied from transmission lines and providing it locally. For example, to compensate an inductive load a shunt capacitor is installed close to the load itself. This allows all reactive power needed by the load to be supplied by the capacitor and not have to be transferred over the transmission lines. This practice alone does not save energy because reactive power does no work, but instead reduces total current flow on the transmission lines allowing them to have smaller current ratings.

Capacitive Vs. Inductive Loads

Stored energy in the magnetic or electric field of a load device, such as a motor or capacitor, causes an offset between the current and the voltage waveforms.

A capacitor is an AC device that stores energy in the form of an electric field. As current is driven through the capacitor, charge build-up causes an opposing voltage to develop across the capacitor. This voltage increases until some maximum dictated by the capacitor structure. In an AC network, the voltage across a capacitor is constantly changing. The capacitor opposes this change, causing the current to lead the voltage in phase. Capacitors are said to "source" reactive power, and thus to cause a leading power factor.

Induction machines are some of the most common types of loads in the electric power system today. These machines use inductors, or large coils of wire to store energy in the form of a magnetic field. When a voltage is initially placed across the coil, the inductor strongly resists this change in current and magnetic field, which causes a time delay for the current to reach its maximum value. This causes the current to lag behind the voltage in phase. Inductors are said to "sink" reactive power, and thus to cause a lagging power factor. Induction generators can source or sink reactive power, and provide a measure of control to system operators over reactive power flow and thus voltage.

Because these devices have opposite effects on the phase angle between voltage and current, they can be used to "cancel out" each other's effects. This usually takes the form of capacitor banks being used to counteract the lagging power factor caused by induction motors.

Reactive Power Control

Transmission connected generators are generally required to support reactive power flow. For example, on the United Kingdom transmission system generators are required by the Grid Code Requirements to supply their rated power between the limits of 0.85 power factor lagging and 0.90 power factor leading at the designated terminals. The system operator will perform switching actions to maintain a secure and economical voltage profile while maintaining a reactive power balance equation:

$$Generator_{MVARs} + System_{gain} + Shunt_{capacitors} = MVAR_{Demand} + Reactive_{losses} + Shunt_{reactors}$$

The 'System gain' is an important source of reactive power in the above power balance equation, which is generated by the capacitive nature of the transmission network itself. By making decisive switching actions in the early morning before the demand increases, the system gain can be maximized early on, helping to secure the system for the whole day.

To balance the equation some pre-fault reactive generator use will be required. Other sources of reactive power that will also be used include shunt capacitors, shunt reactors, Static VAR Compensators and voltage control circuits.

Unbalanced Polyphase Systems

While active power and reactive power are well defined in any system, the definition of apparent power for unbalanced polyphase systems is considered to be one of the most controversial topics in power engineering. Originally, apparent power arose merely as a figure of merit. Major delineations of the concept are attributed to Stanley's *Phenomena of Retardation in the Induction Coil* (1888) and Steinmetz's *Theoretical Elements of Engineering* (1915). However, with the development of three phase power distribution, it became clear that the definition of apparent power and the power factor could not be applied to unbalanced polyphase systems. In 1920, a "Special Joint Committee of the AIEE and the National Electric Light Association" met to resolve the issue. They considered two definitions:

$$pf = \frac{P_a + P_b + P_c}{|S_a| + |S_b| + |S_c|}$$

that is, the quotient of the sums of the active powers for each phase over the sum of the apparent power for each phase.

$$pf = \frac{P_a + P_b + P_c}{|P_a + P_b + P_c + j(Q_a + Q_b + Q_c)|}$$

that is, the quotient of the sums of the active powers for each phase over the magnitude of the sum of the complex powers for each phase.

The 1920 committee found no consensus and the topic continued to dominate discussions. In 1930 another committee formed and once again failed to resolve the question. The transcripts of their discussions are the lengthiest and most controversial ever published by the AIEE (Emanuel, 1993). Further resolution of this debate did not come until the late 1990s.

Real Number Formulas

A perfect resistor stores no energy, so current and voltage are in phase. Therefore, there is no reactive power and $P = S$. Therefore, for a perfect resistor

$$P = S = V_{RMS} I_{RMS} = I_{RMS}^2 R = \frac{V_{RMS}^2}{R}$$

For a perfect capacitor or inductor there is no net power transfer, so all power is reactive. Therefore, for a perfect capacitor or inductor:

$$P = 0$$

$$Q = |S| = V_{RMS}I_{RMS} = I_{RMS}^2 |X| = \frac{V_{RMS}^2}{|X|}$$

Where X is the reactance of the capacitor or inductor.

If X is defined as being positive for an inductor and negative for a capacitor then we can remove the modulus signs from S and X and get

$$Q = I_{RMS}^2 X = \frac{V_{RMS}^2}{X}$$

Instantaneous power is defined as:

$$P_{inst}(t) = V(t)I(t)$$

where v(t) and i(t) are the time varying voltage and current waveforms.

This definition is useful because it applies to all waveforms, whether they are sinusoidal or not. This is particularly useful in power electronics, where nonsinusoidal waveforms are common.

In general, we are interested in the active power averaged over a period of time, whether it is a low frequency line cycle or a high frequency power converter switching period. The simplest way to get that result is to take the integral of the instantaneous calculation over the desired period.

$$P_{avg} = \frac{1}{t_2 - t_1} \int_{t_1}^{t_2} V(t)I(t)dt$$

This method of calculating the average power gives the active power regardless of harmonic content of the waveform. In practical applications, this would be done in the digital domain, where the calculation becomes trivial when compared to the use of rms and phase to determine active power.

$$P_{avg} = \frac{1}{n}\sum_{k=1}^{n} V[k]I[k]$$

Multiple Frequency Systems

Since an RMS value can be calculated for any waveform, apparent power can be calculated from this.

For active power it would at first appear that we would have to calculate many product terms and average all of them. However, if we look at one of these product terms in more detail we come to a very interesting result.

$$A\cos(\omega_1 t + k_1)\cos(\omega_2 t + k_2)$$
$$= \frac{A}{2}\cos\left[(\omega_1 t + k_1) + (\omega_2 t + k_2)\right] + \frac{A}{2}\cos\left[(\omega_1 t + k_1) - (\omega_2 t + k_2)\right]$$
$$= \frac{A}{2}\cos\left[(\omega_1 + \omega_2)t + k_1 + k_2\right] + \frac{A}{2}\cos\left[(\omega_1 - \omega_2)t + k_1 - k_2\right]$$

However, the time average of a function of the form $\cos(\omega t+k)$ is zero provided that ω is nonzero. Therefore, the only product terms that have a nonzero average are those where the frequency of voltage and current match. In other words, it is possible to calculate active (average) power by simply treating each frequency separately and adding up the answers.

Furthermore, if we assume the voltage of the mains supply is a single frequency (which it usually is), this shows that harmonic currents are a bad thing. They will increase the rms current (since there will be non-zero terms added) and therefore apparent power, but they will have no effect on the active power transferred. Hence, harmonic currents will reduce the power factor.

Harmonic currents can be reduced by a filter placed at the input of the device. Typically this will consist of either just a capacitor (relying on parasitic resistance and inductance in the supply) or a capacitor-inductor network. An active power factor correction circuit at the input would generally reduce the harmonic currents further and maintain the power factor closer to unity.

Electricity Generation

Electricity generation is the process of generating electric power from other sources of primary energy. For electric utilities, it is the first process in the delivery of electricity to consumers. The other processes, electricity transmission, distribution, and electrical power storage and recovery using pumped-storage methods are normally carried out by the electric power industry. Electricity is most often generated at a power station by electromechanical generators, primarily driven by heat engines fuelled by combustion or nuclear fission but also by other means such as the kinetic energy of flowing water and wind. Other energy sources include solar photovoltaics and geothermal power.

Turbo generator

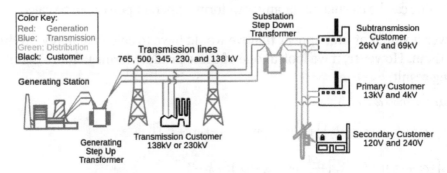

Diagram of an electric power system, generation system in red

History

The fundamental principles of electricity generation were discovered during the 1820s and early 1830s by the British scientist Michael Faraday. This method is still used today: electricity is generated by the movement of a loop of wire, or disc of copper between the poles of a magnet. Central power stations became economically practical with the development of alternating current power transmission, using power transformers to transmit power at high voltage and with low loss. Electricity has been generated at central stations since 1882. The first power plants were run on water power or coal, and today rely mainly on coal, nuclear, natural gas, hydroelectric, wind generators, and petroleum, with a small amount from solar energy, tidal power, and geothermal sources. The use of power-lines and power-poles have been significantly important in the distribution of electricity.

Methods of Generating Electricity

U.S. 2014 Electricity Generation By Type

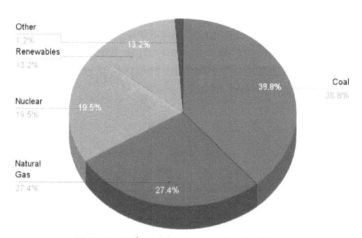

U.S. 2014 Electricity Generation By Type.

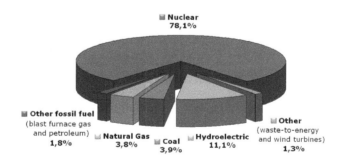

Sources of electricity in France in 2006; nuclear power was the main source.

There are seven fundamental methods of directly transforming other forms of energy into electrical energy.

Static Electricity

Static electricity, form the physical separation and transport of charge (examples: triboelectric effect and lightning). It was the first form discovered and investigated, and the electrostatic generator is still used even in modern devices such as the Van de Graaff generator and MHD generators.

Electromagnetic Induction

In Electromagnetic induction, an electric generator, dynamo or alternator transforms kinetic energy into electricity. This is the most used form for generating electricity and is based on Faraday's law. It can be experimented by rotating a magnet within closed loops of a conducting material (e.g. copper wire). Almost all commercial electrical generation is done using electromagnetic induction, in which mechanical energy forces a generator to rotate.

Turbines

Almost all electrical power on Earth is generated with a turbine, driven by wind, water, steam or burning gas. The turbine drives a generator. There are many different methods of developing mechanical energy, including heat engines, hydro, wind and tidal power. Most electric generation is driven by heat engines. The combustion of fossil fuels supplies most of the heat to these engines, with a significant fraction from nuclear fission and some from renewable sources. The modern steam turbine (invented by Sir Charles Parsons in 1884) currently generates about 80% of the electric power in the world using a variety of heat sources. Power sources include:

- Steam

Large dams such as Three Gorges Dam in China can provide large amounts of hydroelectric power; it has a 22.5 GW capability.

- Water is boiled by coal burned in a thermal power plant, about 41% of all electricity is generated this way.

- Nuclear fission heat created in a nuclear reactor creates steam. Less than 15% of electricity is generated this way.

- Renewables. The steam is generated by Biomass, Solar thermal energy where solar parabolic troughs and solar power towers concentrate sunlight to heat a heat transfer fluid, which is then used to produce steam, or Geothermal power.

- Natural gas: turbines are driven directly by gases produced by combustion. Combined cycle are driven by both steam and natural gas. They generate power by burning natural gas in a gas turbine and use residual heat to generate steam. At least 20% of the worlds electricity is generated by natural gas.

- Small turbines can be powered by Diesel engines. This is used for back up generation, usually at low voltages. Most large power grids also use diesel generators, originally provided as emergency back up for a specific facility such as a hospital, to feed power into the grid during certain circumstances.

- Water Energy is captured from the movement of water. From falling water, the rise and fall of tides or ocean thermal currents. Each driving a water turbine to produce approximately 16% of the world's electricity. The Perth Wave Energy Project is an early production, submerged buoy, electrical power and direct desalination installation supplying power to HMAS Stirling in Western Australia.

- The windmill was a very early wind turbine. In a solar updraft tower wind is artificially produced. Before 2010 less than 2% of the worlds electricity was produced from wind.

Electrochemistry

Large dams such as Hoover Dam can provide large amounts of hydroelectric power; it has 2.07 GW capability.

Electrochemistry is the direct transformation of chemical energy into electricity, as in a battery. Electrochemical electricity generation is important in portable and mobile applications. Currently, most electrochemical power comes from batteries. Primary cells, such as the common zinc-carbon batteries, act as power sources directly, but many types of cells are used as storage systems rather than primary generation systems. Open electrochemical systems, known as fuel cells, can be used to extract power either from natural fuels or from synthesized fuels. Osmotic power is a possibility at places where salt and fresh water merges.

Photovoltaic Effect

The photovoltaic effect is the transformation of light into electrical energy, as in solar cells. Photovoltaic panels convert sunlight directly to electricity. Although sunlight is free and abundant, solar electricity is still usually more expensive to produce than large-scale mechanically generated power due to the cost of the panels. Low-efficiency silicon solar cells have been decreasing in cost and multijunction cells with close to 30% conversion efficiency are now commercially available. Over

40% efficiency has been demonstrated in experimental systems. Until recently, photovoltaics were most commonly used in remote sites where there is no access to a commercial power grid, or as a supplemental electricity source for individual homes and businesses. Recent advances in manufacturing efficiency and photovoltaic technology, combined with subsidies driven by environmental concerns, have dramatically accelerated the deployment of solar panels. Installed capacity is growing by 40% per year led by increases in Germany, Japan, and the United States.

Thermoelectric Effect

Thermoelectric effect is the direct conversion of temperature differences to electricity, as in thermocouples, thermopiles, and thermionic converters.

A coal-fired power plant in Laughlin, Nevada U.S.A. Owners of this plant ceased operations after declining to invest in pollution control equipment to comply with pollution regulations.

Piezoelectric Effect

The Piezoelectric effect generates electricity from the mechanical strain of electrically anisotropic molecules or crystals. Researchers at the United States Department of Energy's Lawrence Berkeley National Laboratory have developed a piezoelectric generator sufficient to operate a liquid crystal display using thin films of M13 bacteriophage. Piezoelectric devices are used for power generation from mechanical strain, particularly in power harvesting.

Nuclear Transformation

Nuclear transformation is the creation and acceleration of charged particles (examples: betavoltaics or alpha particle emission). The direct conversion of nuclear potential energy to electricity by beta decay is used only on a small scale. In a full-size nuclear power plant, the heat of a nuclear reaction is used to run a heat engine. This drives a generator, which converts mechanical energy into electricity by magnetic induction. Betavoltaics are a type of solid-state power generator which produces electricity from radioactive decay. Fluid-based magnetohydrodynamic (MHD) power generation has been studied as a method for extracting electrical power from nuclear reactors and also from more conventional fuel combustion systems.

Wind turbines usually provide electrical generation in conjunction with other methods of producing power.

Economics of Generation and Production of Electricity

The selection of electricity production modes and their economic viability varies in accordance with demand and region. The economics vary considerably around the world, resulting in widespread selling prices, e.g. the price in Venezuela is 3 cents per kWh while in Denmark it is 40 cents per kWh. Hydroelectric plants, nuclear power plants, thermal power plants and renewable sources have their own pros and cons, and selection is based upon the local power requirement and the fluctuations in demand. All power grids have varying loads on them but the daily minimum is the base load, supplied by plants which run continuously. Nuclear, coal, oil and gas plants can supply base load.

Thermal energy is economical in areas of high industrial density, as the high demand cannot be met by renewable sources. The effect of localized pollution is also minimized as industries are usually located away from residential areas. These plants can also withstand variation in load and consumption by adding more units or temporarily decreasing the production of some units. Nuclear power plants can produce a huge amount of power from a single unit. However, recent disasters in Japan have raised concerns over the safety of nuclear power, and the capital cost of nuclear plants is very high. Hydroelectric power plants are located in areas where the potential energy from falling water can be harnessed for moving turbines and the generation of power. It is not an economically viable source of production where the load varies too much during the annual production cycle and the ability to store the flow of water is limited.

Due to advancements in technology, and with mass production, renewable sources other than hydroelectricity (solar power, wind energy, tidal power, etc.) experienced decreases in cost of production, and the energy is now in many cases cost-comparative with fossil fuels. Many governments around the world provide subsidies to offset the higher cost of any new power production, and to make the installation of renewable energy systems economically feasible. However, their use is frequently limited by their intermittent nature. If natural gas prices are below $3 per million British thermal units, generating electricity from natural gas is cheaper than generating power by burning coal.

Production

The production of electricity in 2009 was 20,053TWh. Sources of electricity were fossil fuels 67%, renewable energy 16% (mainly hydroelectric, wind, solar and biomass), and nuclear power 13%, and other sources were 3%. The majority of fossil fuel usage for the generation of electricity was coal and gas. Oil was 5.5%, as it is the most expensive common commodity used to produce electrical energy. Ninety-two percent of renewable energy was hydroelectric followed by wind at 6% and geothermal at 1.8%. Solar photovoltaic was 0.06%, and solar thermal was 0.004%. Data are from OECD 2011-12 Factbook (2009 data).

Source of Electricity (World total year 2008)							
-	Coal	Oil	Natural Gas	Nuclear	Renewables	other	Total
Average electric power (TWh/year)	8,263	1,111	4,301	2,731	3,288	568	20,261
Average electric power (GW)	942.6	126.7	490.7	311.6	375.1	64.8	2311.4
Proportion	41%	5%	21%	13%	16%	3%	100%

Data Source IEA/OECD

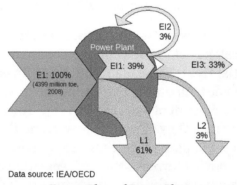

Data source: IEA/OECD

Energy Flow of Power Plant

Total energy consumed at all power plants for the generation of electricity was 4,398,768 ktoe (kilo ton of oil equivalent) which was 36% of the total for primary energy sources (TPES) of 2008. Electricity output (gross) was 1,735,579 ktoe (20,185 TWh), efficiency was 39%, and the balance of 61% was generated heat. A small part (145,141 ktoe, which was 3% of the input total) of the heat was utilized at co-generation heat and power plants. The in-house consumption of electricity and power transmission losses were 289,681 ktoe. The amount supplied to the final consumer was 1,445,285 ktoe (16,430 TWh) which was 33% of the total energy consumed at power plants and heat and power co-generation (CHP) plants.

Historical Results of Production of Electricity

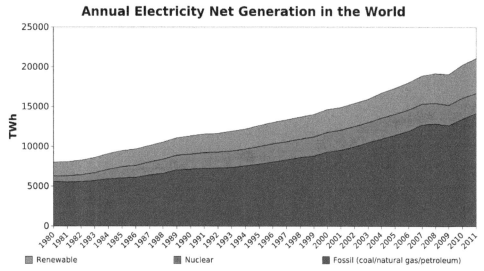

Annual Electricity Net Generation in the World

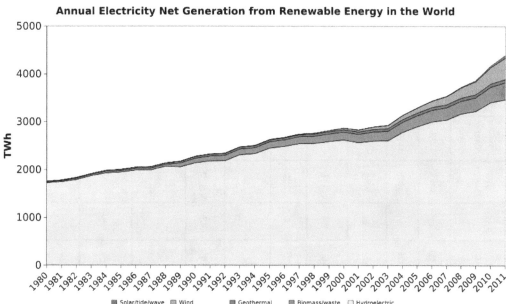

Annual Electricity Net Generation from Renewable Energy in the World

Production by Country

The United States has long been the largest producer and consumer of electricity, with a global share in 2005 of at least 25%, followed by China, Japan, Russia, and India. As of Jan-2010, total electricity generation for the 2 largest generators was as follows: USA: 3992 billion kWh (3992 TWh) and China: 3715 billion kWh (3715 TWh).

List of Countries with Source of Electricity 2008

Data source of values (electric power generated) is IEA/OECD. Listed countries are top 20 by population or top 20 by GDP (PPP) and Saudi Arabia based on CIA World Factbook 2009.

Composition of Electricity by Resource (TWh per year 2008)

Country's electricity sector	Fossil Fuel					Nuclear	rank	Renewable								Bio other*	total	rank
	Coal	Oil	Gas	sub total	rank	Nuclear	rank	Hydro	Geo Thermal	Solar PV*	Solar Thermal	Wind	Tide	sub total	rank	Bio other*	total	rank
World total	8,263	1,111	4,301	13,675	-	2,731	-	3,288	65	12	0.9	219	0.5	3,584	-	271	20,261	-
Proportion	41%	5.5%	21%	67%	-	13%	-	16%	0.3%	0.06%	0.004%	1.1%	0.003%	18%	-	1.3%	100%	-
China	2,733	23	31	2,788	2	68	8	585	-	0.2	-	13	-	598	1	2.4	3,457	2
India	569	34	82	685	5	15	12	114	-	0.02	-	14	-	128.02	6	2.0	830	5
USA	2,133	58	1011	3,101	1	838	1	282	17	1.6	0.88	56	-	357	4	73	4,369	1
Indonesia	61	43	25	130	19	-	-	12	8.3	-	-	-	-	20	17	-	149	20
Brazil	13	18	29	59	23	14	13	370	-	-	-	0.6	-	370	3	20	463	9
Pakistan	0.1	32	30	62	22	1.6	16	28	-	-	-	-	-	28	14	-	92	24
Bangladesh	0.6	1.7	31	33	27	-	-	1.5	-	-	-	-	-	1.5	29	-	35	27
Nigeria	-	3.1	12	15	28	-	-	5.7	-	-	-	-	-	5.7	25	-	21	28
Russia	197	16	495	708	4	163	4	167	0.5	-	-	0.01	-	167	5	2.5	1,040	4
Japan	288	139	283	711	3	258	3	83	2.8	2.3	-	2.6	-	91	7	22	1,082	3

Composition of Electricity by Resource (TWh per year 2008)

Country's electricity sector	Fossil Fuel					Nuclear		Renewable								Bio other*	total	rank
	Coal	Oil	Gas	sub total	rank	Nuclear	rank	Hydro	Geo Thermal	Solar PV*	Solar Thermal	Wind	Tide	sub total	rank			
Mexico	21	49	131	202	13	9.8	14	39	7.1	0.01	-	0.3	-	47	12	0.8	259	14
Philippines	16	4.9	20	40	26	-	-	9.8	11	0.001	-	0.1	-	21	16	-	61	26
Vietnam	15	1.6	30	47	25	-	-	26	-	-	-	-	-	26	15	-	73	25
Ethiopia	-	0.5	-	0.5	29	-	-	3.3	0.01	-	-	-	-	3.3	28	-	3.8	30
Egypt	-	26	90	115	20	-	-	15	-	-	-	0.9	-	16	20	-	131	22
Germany	291	9.2	88	388	6	148	6	27	0.02	4.4	-	41	-	72	9	29	637	7
Turkey	58	7.5	99	164	16	-	-	33	0.16	-	-	0.85	-	34	13	0.22	198	19
DR Congo	-	0.02	0.03	0.05	30	-	-	7.5	-	-	-	-	-	7.5	22	-	7.5	29
Iran	0.4	36	173	209	11	-	-	5.0	-	-	-	0.20	-	5.2	26	-	215	17
Thailand	32	1.7	102	135	18	-	-	7.1	0.002	0.003	-	-	-	7.1	23	4.8	147	21
France	27	5.8	22	55	24	439	2	68	-	0.04	-	5.7	0.51	75	8	5.9	575	8

Composition of Electricity by Resource (TWh per year 2008)

Country's electricity sector	Fossil Fuel					Nuclear		Renewable								Bio other*	Total	rank
	Coal	Oil	Gas	sub total	rank	Nuclear	rank	Hydro	Geo Thermal	Solar PV	Solar Thermal	Wind	Tide	sub total	rank	Bio other	Total	rank
UK	127	6.1	177	310	7	52	10	9.3	-	0.02	-	7.1	-	16	18	11	389	11
Italy	49	31	173	253	9	-	-	47	5.5	0.2	-	4.9	-	58	11	8.6	319	12
South Korea	192	15	81	288	8	151	5	5.6	-	0.3	-	0.4	-	6.3	24	0.7	446	10
Spain	50	18	122	190	14	59	9	26	-	2.6	0.02	32	-	61	10	4.3	314	13
Canada	112	9.8	41	162	17	94	7	383	-	0.03	-	3.8	0.03	386	2	8.5	651	6
Saudi Arabia	-	116	88	204	12	-	-	-	-	-	-	-	-	-	-	-	204	18
Taiwan	125	14	46	186	15	41	11	7.8	-	0.004	-	0.6	-	8.4	21	3.5	238	16
Australia	198	2.8	39	239	10	-	-	12	-	0.2	0.004	3.9	-	16	19	2.2	257	15
Netherlands	27	2.1	63	92	21	4.2	15	0.1	-	0.04	-	4.3	-	4.4	27	6.8	108	23

Solar PV* is Photovoltaics Bio other* = 198TWh (Biomass) + 69TWh (Waste) + 4TWh (other)

Environmental Concerns

Variations between countries generating electrical power affect concerns about the environment. In France only 10% of electricity is generated from fossil fuels, the US is higher at 70% and China is at 80%. The cleanliness of electricity depends on its source. Most scientists agree that emissions of pollutants and greenhouse gases from fossil fuel-based electricity generation account for a significant portion of world greenhouse gas emissions; in the United States, electricity generation accounts for nearly 40% of emissions, the largest of any source. Transportation emissions are close behind, contributing about one-third of U.S. production of carbon dioxide. In the United States, fossil fuel combustion for electric power generation is responsible for 65% of all emissions of sulfur dioxide, the main component of acid rain. Electricity generation is the fourth highest combined source of NOx, carbon monoxide, and particulate matter in the US. In July 2011, the UK parliament tabled a motion that "levels of (carbon) emissions from nuclear power were approximately three times lower per kilowatt hour than those of solar, four times lower than clean coal and 36 times lower than conventional coal".

Power Station

The Athlone Power Station in Cape Town, South Africa.

Hydroelectric power station at Gabčíkovo Dam, Slovakia.

A power station, also referred to as a generating station, power plant, powerhouse, or generating plant, is an industrial facility for the generation of electric power. Most power stations contain one or more generators, a rotating machine that converts mechanical power into electrical power. The relative motion between a magnetic field and a conductor creates an electrical current. The energy source harnessed to turn the generator varies widely. Most power stations in the world burn fossil fuels such as coal, oil, and natural gas to generate electricity. Others use nuclear power, but there is an increasing use of cleaner renewable sources such as solar, wind, wave and hydroelectric.

History

In 1868 a hydro electric power station was designed and built by Lord Armstrong at Cragside, England. It used water from lakes on his estate to power Siemens dynamos. The electricity supplied power to lights, heating, produced hot water, ran an elevator as well as labor-saving devices and farm buildings.

In the early 1870s Belgian inventor Zénobe Gramme invented a generator powerful enough to produce power on a commercial scale for industry.

In the autumn of 1882, a central station providing public power was built in Godalming, England. It was proposed after the town failed to reach an agreement on the rate charged by the gas company, so the town council decided to use electricity. It used hydroelectric power that was used to street and household lighting. The system was not a commercial success and the town reverted to gas.

In 1882 a public power station, the Edison Electric Light Station, was built in London, was a project of Thomas Edison organized by Edward Johnson. A Babcock & Wilcox boiler powered a 125-horse-power steam engine that drove a 27-ton generator. This supplied electricity to premises in the area that could be reached through the culverts of the viaduct without digging up the road, which was the monopoly of the gas companies. The customers included the City Temple and the Old Bailey. Another important customer was the Telegraph Office of the General Post Office, but this could not be reached though the culverts. Johnson arranged for the supply cable to be run overhead, via Holborn Tavern and Newgate.

In September 1882 in New York, the Pearl Street Station was established by Edison to provide electric lighting in the lower Manhattan Island area. The station ran until destroyed by fire in 1890. The station used reciprocating steam engines to turn direct-current generators. Because of the DC distribution, the service area was small, limited by voltage drop in the feeders. The War of Currents eventually resolved in favor of AC distribution and utilization, although some DC systems persisted to the end of the 20th century. DC systems with a service radius of a mile (kilometer) or so were necessarily smaller, less efficient of fuel consumption, and more labor-intensive to operate than much larger central AC generating stations.

AC systems used a wide range of frequencies depending on the type of load; lighting load using higher frequencies, and traction systems and heavy motor load systems preferring lower frequencies. The economics of central station generation improved greatly when unified light and power systems, operating at a common frequency, were developed. The same generating plant that fed large industrial loads during the day, could feed commuter railway systems during rush hour and then serve lighting load in the evening, thus improving the system load factor and reducing the cost of electrical energy overall. Many exceptions existed, generating stations were dedicated to power or light by the choice of frequency, and rotating frequency changers and rotating converters were particularly common to feed electric railway systems from the general lighting and power network.

Throughout the first few decades of the 20th century central stations became larger, using higher steam pressures to provide greater efficiency, and relying on interconnections of multiple generating stations to improve reliability and cost. High-voltage AC transmission allowed hydroelectric power to be conveniently moved from distant waterfalls to city markets. The advent of the steam turbine in central station service, around 1906, allowed great expansion of generating capacity. Generators were no longer limited by the power transmission of belts or the relatively slow speed of reciprocating engines, and could grow to enormous sizes. For example, Sebastian Ziani de Ferranti planned what would have been the largest reciprocating steam engine ever built for a proposed new central station, but scrapped the plans when turbines became available in the necessary size. Building power systems out of central stations required combinations of engineering skill and financial acumen in equal measure. Pioneers of central station generation include George Westinghouse and Samuel Insull in the United States, Ferranti and Charles Hesterman Merz in UK, and many others.

Thermal Power Stations

Rotor of a modern steam turbine, used in a power station.

In thermal power stations, mechanical power is produced by a heat engine that transforms thermal energy, often from combustion of a fuel, into rotational energy. Most thermal power stations produce steam, so they are sometimes called steam power stations. Not all thermal energy can be transformed into mechanical power, according to the second law of thermodynamics; therefore, there is always heat lost to the environment. If this loss is employed as useful heat, for industrial processes or district heating, the power plant is referred to as a cogeneration power plant or CHP (combined heat-and-power) plant. In countries where district heating is common, there are dedicated heat plants called heat-only boiler stations. An important class of power stations in the Middle East uses by-product heat for the desalination of water.

The efficiency of a thermal power cycle is limited by the maximum working fluid temperature produced. The efficiency is not directly a function of the fuel used. For the same steam conditions, coal-, nuclear- and gas power plants all have the same theoretical efficiency. Overall, if a system is on constantly (base load) it will be more efficient than one that is used intermittently (peak load). Steam turbines generally operate at higher efficiency when operated at full capacity.

Besides use of reject heat for process or district heating, one way to improve overall efficiency of a power plant is to combine two different thermodynamic cycles in a combined cycle plant. Most commonly, exhaust gases from a gas turbine are used to generate steam for a boiler and a steam turbine. The combination of a "top" cycle and a "bottom" cycle produces higher overall efficiency than either cycle can attain alone.

Classification

St. Clair Power Plant, a large coal-fired generating station in Michigan, United States.

Ikata Nuclear Power Plant, Japan.

Nesjavellir Geothermal Power Station, Iceland.

By Heat Source

- Fossil-fuel power stations may also use a steam turbine generator or in the case of natural gas-fired plants may use a combustion turbine. A coal-fired power station produces heat by burning coal in a steam boiler. The steam drives a steam turbine and generator that then produces electricity. The waste products of combustion include ash, sulphur dioxide, nitrogen oxides and carbon dioxide. Some of the gases can be removed from the waste stream to reduce pollution.

- Nuclear power plants use a nuclear reactor's heat that is transferred to steam which then operates a steam turbine and generator. About 20 percent of electric generation in the USA is produced by nuclear power plants.

- Geothermal power plants use steam extracted from hot underground rocks. These rocks are heated by the decay of radioactive material in the Earth's crust.

- Biomass-fuelled power plants may be fuelled by waste from sugar cane, municipal solid waste, landfill methane, or other forms of biomass.

- In integrated steel mills, blast furnace exhaust gas is a low-cost, although low-energy-density, fuel.

- Waste heat from industrial processes is occasionally concentrated enough to use for power generation, usually in a steam boiler and turbine.

- Solar thermal electric plants use sunlight to boil water and produce steam which turns the generator.

By Prime Mover

- Steam turbine plants use the dynamic pressure generated by expanding steam to turn the blades of a turbine. Almost all large non-hydro plants use this system. About 90 percent of all electric power produced in the world is through use of steam turbines.

- Gas turbine plants use the dynamic pressure from flowing gases (air and combustion products) to directly operate the turbine. Natural-gas fuelled (and oil fueled) combustion turbine plants can start rapidly and so are used to supply "peak" energy during periods of high demand, though at higher cost than base-loaded plants. These may be comparatively small units, and sometimes completely unmanned, being remotely operated. This type was pioneered by the UK, Princetown being the world's first, commissioned in 1959.

- Combined cycle plants have both a gas turbine fired by natural gas, and a steam boiler and steam turbine which use the hot exhaust gas from the gas turbine to produce electricity. This greatly increases the overall efficiency of the plant, and many new baseload power plants are combined cycle plants fired by natural gas.

- Internal combustion reciprocating engines are used to provide power for isolated communities and are frequently used for small cogeneration plants. Hospitals, office buildings, industrial plants, and other critical facilities also use them to provide backup power in case of a power outage. These are usually fuelled by diesel oil, heavy oil, natural gas, and landfill gas.

- Microturbines, Stirling engine and internal combustion reciprocating engines are low-cost solutions for using opportunity fuels, such as landfill gas, digester gas from water treatment plants and waste gas from oil production.

By Duty

Power plants that can be dispatched (scheduled) to provide energy to a system include:

- Base load power plants run nearly continually to provide that component of system load that doesn't vary during a day or week. Baseload plants can be highly optimized for low fuel cost, but may not start or stop quickly during changes in system load. Examples of base-load plants would include large modern coal-fired and nuclear generating stations, or hydro plants with a predictable supply of water.

- Peaking power plants meet the daily peak load, which may only be for one or two hours each day. While their incremental operating cost is always higher than base load plants, they are required to ensure security of the system during load peaks. Peaking plants include simple cycle gas turbines and sometimes reciprocating internal combustion engines, which can be started up rapidly when system peaks are predicted. Hydroelectric plants may also be designed for peaking use.

- Load following power plants can economically follow the variations in the daily and weekly load, at lower cost than peaking plants and with more flexibility than baseload plants.

Non-dispatchable plants include such sources as wind and solar energy; while their long-term contribution to system energy supply is predictable, on a short-term (daily or hourly) base their energy must be used as available since generation cannot be deferred. Contractual arrangements ("take or pay") with independent power producers or system interconnections to other networks may be effectively non-dispatchable.

Cooling Towers

Cooling towers showing evaporating water at Ratcliffe-on-Soar Power Station, United Kingdom.

"Camouflaged" natural draft wet cooling tower

All thermal power plants produce waste heat energy as a byproduct of the useful electrical energy produced. The amount of waste heat energy equals or exceeds the amount of energy converted into useful electricity. Gas-fired power plants can achieve as much as 65 percent conversion efficiency, while coal and oil plants achieve around 30 to 49 percent. The waste heat produces a temperature rise in the atmosphere, which is small compared to that produced by greenhouse-gas emissions from the same power plant. Natural draft wet cooling towers at many nuclear power plants and large fossil fuel-fired power plants use large hyperboloid chimney-like structures that release the waste heat to the ambient atmosphere by the evaporation of water.

However, the mechanical induced-draft or forced-draft wet cooling towers in many large thermal power plants, nuclear power plants, fossil-fired power plants, petroleum refineries, petrochemical plants, geothermal, biomass and waste-to-energy plants use fans to provide air movement upward through downcoming water, and are not hyperboloid chimney-like structures. The induced or forced-draft cooling towers are typically rectangular, box-like structures filled with a material that enhances the mixing of the upflowing air and the downflowing water.

In areas with restricted water use, a dry cooling tower or directly air-cooled radiators may be necessary, since the cost or environmental consequences of obtaining make-up water for evaporative cooling would be prohibitive. These coolers have lower efficiency and higher energy consumption to drive fans, compared to a typical wet, evaporative cooling tower.

Once-through Cooling Systems

Electric companies often prefer to use cooling water from the ocean, a lake, or a river, or a cooling pond, instead of a cooling tower. This single pass or once-through cooling system can save the cost of a cooling tower and may have lower energy costs for pumping cooling water through the plant's heat exchangers. However, the waste heat can cause thermal pollution as the water is discharged. Power plants using natural bodies of water for cooling are designed with mechanisms such as fish screens, to limit intake of organisms into the cooling machinery. These screens are only partially effective and as a result billions of fish and other aquatic organisms are killed by power plants each year. For example, the cooling system at the Indian Point Energy Center in New York kills over a billion fish eggs and larvae annually.

A further environmental impact is that aquatic organisms which adapt to the warmer discharge water may be injured if the plant shuts down in cold weather.

Water consumption by power stations is a developing issue.

In recent years, recycled wastewater, or grey water, has been used in cooling towers. The Calpine Riverside and the Calpine Fox power stations in Wisconsin as well as the Calpine Mankato power station in Minnesota are among these facilities.

Power from Renewable Energy

Power stations can also generate electrical energy from renewable energy sources.

Hydroelectric Power Station

Three Gorges Dam, Hubei, China.

In a hydroelectric power station water flows though turbines using hydropower to generate hydroelectricity. Power is captured from the gravitational force of water falling through penstocks to water turbines connected to generators. The amount of power available is a combination of height and flow. A wide range of Dams may be built to raise the water level, and create a lake for storing water. Hydropower is produced in 150 countries, with the Asia-Pacific region generating 32 percent of global hydropower in 2010. China is the largest hydroelectricity producer, with 721 terawatt-hours of production in 2010, representing around 17 percent of domestic electricity use.

Pumped Storage

A pumped-storage is a reversible hydroelectric power plant. They are a net consumer of energy but can be used for storage to smooth peaks and troughs in overall electricity demand. Pumped storage plants typically use "spare" electricity during off peak periods to pump freshwater or saltwater from a lower reservoir to an upper reservoir. Because the pumping takes place "off peak", electricity is typically cheaper than at peak times. This is because power sources such as coal-fired, solar and wind are not switched off and remain in service even when demand is low. During hours of peak demand, when the electricity price is high, the water pumped to the upper reservoir is allowed to flow back to the lower reservoir through a water turbine connected to an electricity generator. Unlike coal power stations, which can take more than 12 hours to start up from cold, the hydroelectric plant can be brought into service in a few minutes, ideal to meet a peak load demand. Two substantial pumped storage schemes are in South Africa, Palmiet Pumped Storage Scheme and another in the Drakensberg, Ingula Pumped Storage Scheme.

Solar

Nellis Solar Power Plant in Nevada, United States.

Solar energy can be turned into electricity either directly in solar cells, or in a concentrating solar power plant by focusing the light to run a heat engine.

A solar photovoltaic power plant converts sunlight into direct current electricity using the photoelectric effect. Inverters change the direct current into alternating current for connection to the electrical grid. This type of plant does not use rotating machines for energy conversion.

Solar thermal power plants are another type of solar power plant. They use either parabolic troughs or heliostats to direct sunlight onto a pipe containing a heat transfer fluid, such as oil. The heated oil is then used to boil water into steam, which turns a turbine that drives an electrical generator. The central tower type of solar thermal power plant uses hundreds or thousands of mirrors, depending on size, to direct sunlight onto a receiver on top of a tower. Again, the heat is used to produce steam to turn turbines that drive electrical generators.

Wind

Wind turbines in Texas, United States.

Wind turbines can be used to generate electricity in areas with strong, steady winds, sometimes offshore. Many different designs have been used in the past, but almost all modern turbines being produced today use a three-bladed, upwind design. Grid-connected wind turbines now being built are much larger than the units installed during the 1970s. They thus produce power more cheaply and reliably than earlier models. With larger turbines (on the order of one megawatt), the blades move more slowly than older, smaller, units, which makes them less visually distracting and safer for birds.

Marine

Marine energy or marine power (also sometimes referred to as ocean energy or ocean power) refers to the energy carried by ocean waves, tides, salinity, and ocean temperature differences. The movement of water in the world's oceans creates a vast store of kinetic energy, or energy in motion. This energy can be harnessed to generate electricity to power homes, transport and industries.

The term marine energy encompasses both wave power — power from surface waves, and tidal power — obtained from the kinetic energy of large bodies of moving water. Offshore wind power is not a form of marine energy, as wind power is derived from the wind, even if the wind turbines are placed over water.

The oceans have a tremendous amount of energy and are close to many if not most concentrated populations. Ocean energy has the potential of providing a substantial amount of new renewable energy around the world.

Osmosis

Osmotic Power Prototype at Tofte (Hurum), Norway

Salinity gradient energy is called pressure-retarded osmosis. In this method, seawater is pumped into a pressure chamber that is at a pressure lower than the difference between the pressures of saline water and fresh water. Freshwater is also pumped into the pressure chamber through a membrane, which increases both the volume and pressure of the chamber. As the pressure differences are compensated, a turbine is spun creating energy. This method is being specifically studied by the Norwegian utility Statkraft, which has calculated that up to 25 TWh/yr would be available from this process in Norway. Statkraft has built the world's first prototype osmotic power plant on the Oslo fiord which was opened on November 24, 2009.

Biomass

Metz biomass power station

Biomass energy can be produced from combustion of waste green material to heat water into steam and drive a steam turbine. Bioenergy can also be processed through a range of temperatures and pressures in gasification, pyrolysis or torrefaction reactions. Depending on the desired end product, these reactions create more energy-dense products (syngas, wood pellets, biocoal) that can then be fed into an accompanying engine to produce electricity at a much lower emission rate when compared with open burning.

Storage Power Stations

It is possible to store energy and produce the electricity at a later time like in Pumped-storage hydroelectricity, Thermal energy storage, Flywheel energy storage, Battery storage power station and so on.

Typical Power Output

The power generated by a power station is measured in multiples of the watt, typically megawatts (10^6 watts) or gigawatts (10^9 watts). Power stations vary greatly in capacity depending on the type of power plant and on historical, geographical and economic factors. The following examples offer a sense of the scale.

Many of the largest operational onshore wind farms are located in the USA. As of 2011, the Roscoe Wind Farm is the second largest onshore wind farm in the world, producing 781.5 MW of power, followed by the Horse Hollow Wind Energy Center (735.5 MW). As of July 2013, the London Array in United Kingdom is the largest offshore wind farm in the world at 630 MW, followed by Thanet Offshore Wind Project in United Kingdom at 300 MW.

As of 2015, the largest photovoltaic (PV) power plants in the world are led by Longyangxia Dam Solar Park in China, rated at 850 megawatts.

Solar thermal power stations in the U.S. have the following output:

The country's largest solar facility at Kramer Junction has an output of 354 MW

The Blythe Solar Power Project planned production is estimated at 485 MW

Aerial view of the Three Mile Island Nuclear Generating Station, USA.

Large coal-fired, nuclear, and hydroelectric power stations can generate hundreds of megawatts to multiple gigawatts. Some examples:

The Three Mile Island Nuclear Generating Station in the USA has a rated capacity of 802 megawatts.

The coal-fired Ratcliffe-on-Soar Power Station in the UK has a rated capacity of 2 gigawatts.

The Aswan Dam hydro-electric plant in Egypt has a capacity of 2.1 gigawatts.

The Three Gorges Dam hydro-electric plant in China has a capacity of 22.5 gigawatts.

Gas turbine power plants can generate tens to hundreds of megawatts. Some examples:

The Indian Queens simple-cycle peaking power station in Cornwall UK, with a single gas turbine is rated 140 megawatts.

The Medway Power Station, a combined-cycle power station in Kent, UK with two gas turbines and one steam turbine, is rated 700 megawatts.

The rated capacity of a power station is nearly the maximum electrical power that that power station can produce. Some power plants are run at almost exactly their rated capacity all the time, as a non-load-following base load power plant, except at times of scheduled or unscheduled maintenance.

However, many power plants usually produce much less power than their rated capacity.

In some cases a power plant produces much less power than its rated capacity because it uses an intermittent energy source. Operators try to pull maximum available power from such power plants, because their marginal cost is practically zero, but the available power varies widely—in particular, it may be zero during heavy storms at night.

In some cases operators deliberately produce less power for economic reasons. The cost of fuel to run a load following power plant may be relatively high, and the cost of fuel to run a peaking power plant is even higher—they have relatively high marginal costs. Operators keep power plants turned off ("operational reserve") or running at minimum fuel consumption ("spinning reserve") most of the time. Operators feed more fuel into load following power plants only when the demand rises above what lower-cost plants (i.e., intermittent and base load plants) can produce, and then feed more fuel into peaking power plants only when the demand rises faster than the load following power plants can follow.

Operations

Control room of a power plant

The term Power station is generally limited to those able to be despatched by a system operator (i.e. the system operator can, by one means or another, alter the planned output of the generating facility).

The power station operator has several duties in the electricity-generating facility. Operators are responsible for the safety of the work crews that frequently do repairs on the mechanical and electrical equipment. They maintain the equipment with periodic inspections and log temperatures, pressures and other important information at regular intervals. Operators are responsible for starting and stopping the generators depending on need. They are able to synchronize and adjust the voltage output of the added generation with the running electrical system, without upsetting the system. They must know the electrical and mechanical systems in order to troubleshoot solve/fix problems in the facility and add to the reliability of the facility. Operators must be able to respond to an emergency and know the procedures in place to deal with it.

References

- Croft, Terrell; Summers, Wilford I. (1987). American Electricians' Handbook (Eleventh ed.). New York: McGraw Hill. ISBN 0-07-013932-6.

- Fink, Donald G.; Beaty, H. Wayne (1978). Standard Handbook for Electrical Engineers (Eleventh ed.). New York: McGraw Hill. ISBN 0-07-020974-X.

- Howard M. Berlin, Frank C. Getz, Principles of Electronic Instrumentation and Measurement, p. 37, Merrill Pub. Co., 1988 ISBN 0-675-20449-6.

- K. S. Suresh Kumar, Electric Circuit Analysis, Pearson Education India, 2013, ISBN 9332514100, section 1.2.3 "'Current intensity' is usually referred to as 'current' itself."

- N. N. Bhargava; D. C. Kulshreshtha (1983). Basic Electronics & Linear Circuits. Tata McGraw-Hill Education. p. 90. ISBN 978-0-07-451965-3.

- Andrew J. Robinson; Lynn Snyder-Mackler (2007). Clinical Electrophysiology: Electrotherapy and Electrophysiologic Testing (3rd ed.). Lippincott Williams & Wilkins. p. 10. ISBN 978-0-7817-4484-3.

- Rudolf Holze, Experimental Electrochemistry: A Laboratory Textbook, page 44, John Wiley & Sons, 2009 ISBN 3527310983.

- N. N. Bhargava & D. C. Kulshreshtha (1983). Basic Electronics & Linear Circuits. Tata McGraw-Hill Education. p. 90. ISBN 978-0-07-451965-3.

- Hughes, Thomas P. (1993). Networks of Power: Electrification in Western Society, 1880-1930. Baltimore: The Johns Hopkins University Press. p. 96. ISBN 0-8018-2873-2. Retrieved Sep 9, 2009.

- McNichol, Tom (2006). AC/DC: the savage tale of the first standards war. John Wiley and Sons. p. 80. ISBN 978-0-7879-8267-6.

- Hughes, Thomas P. (1993). Networks of Power: Electrification in Western Society, 1880-1930. Baltimore: The Johns Hopkins University Press. p. 98. ISBN 0-8018-2873-2

- Kreith, l Frank; D. Yogi Goswami (2005). The CRC Handbook Of Mechanical Engineering, 2nd Ed. CRC Press. pp. 5.5–5.6. ISBN 0849308666.

- Glisson, Tildon H. (2011). Introduction to Circuit Analysis and Design. USA: Springer. pp. 114–116. ISBN 9048194423.

- Eccles, William J. (2011). Pragmatic Electrical Engineering: Fundamentals. Morgan & Claypool Publishers. pp. 4–5. ISBN 1608456684.

- O'Malley, John (1992). Schaum's Outline of Basic Circuit Analysis, 2nd Ed. McGraw Hill Professional. pp. 2–4. ISBN 0070478244.

- Kumar, K. S. Suresh (2008). Electric Circuits & Networks. Pearson Education India. pp. 26–28. ISBN 8131713903.

- Glover, J. Duncan; Mulukutla S. Sarma; Thomas Jeffrey Overbye (2011). Power System Analysis and Design, 5th Ed. Cengage Learning. pp. 53–54. ISBN 1111425779.

- McNeil, Ian (1996). An Encyclopaedia of the History of Technology ([New ed.]. ed.). London: Routledge. p. 369. ISBN 978-0-415-14792-7.

- Wiser, Wendell H. (2000). Energy resources: occurrence, production, conversion, use. Birkhäuser. p. 190. ISBN 978-0-387-98744-6.

- Thomas C. Elliott, Kao Chen, Robert Swanekamp (coauthors) (1997). Standard Handbook of Powerplant Engineering (2nd ed.). McGraw-Hill Professional. ISBN 0-07-019435-1

- British Electricity International (1991). Modern Power Station Practice: incorporating modern power system practice (3rd Edition (12 volume set) ed.). Pergamon. ISBN 0-08-040510-X.

Measurement Units used in Electrical Engineering

Volt is the unit used in electrical engineering, and ampere is the SI unit of electric current. The other measurement units elucidated in this text are ohm, farad and henry. This section is an overview of the subject matter incorporating all the major aspects of the measurement units of electrical engineering.

Volt

The volt (symbol: V) is the derived unit for electric potential, electric potential difference (voltage), and electromotive force. The volt is named in honour of the Italian physicist Alessandro Volta (1745–1827), who invented the voltaic pile, possibly the first chemical battery.

Description

The volt is a measure of electric potential. Electrical potential is a type of potential energy, and refers to the energy that could be released if electric current is allowed to flow. An analogy is that a suspended object (a mass) that is said to have gravitational potential energy (as a result of a gravitational field), which is the amount of energy that would be released if the object was allowed to fall.

In alternating current, the voltages increase, decrease and change direction at regular intervals. As a result, voltage for alternating current almost never refers to the voltage at a particular instant, but instead is the root mean square (RMS) voltage, which is a way of defining an effective voltage when calculating power (RMS voltage × RMS current). In most cases, the fact that a voltage is an RMS voltage is not explicitly stated, but assumed.

Definition

One volt is defined as the difference in electric potential between two points of a conducting wire when an electric current of one ampere dissipates one watt of power between those points. It is also equal to the potential difference between two parallel, infinite planes spaced 1 meter apart that create an electric field of 1 newton per coulomb. Additionally, it is the potential difference between two points that will impart one joule of energy per coulomb of charge that passes through it. It can be expressed in terms of SI base units (m, kg, s, and A) as

$$V = \frac{\text{potential energy}}{\text{charge}} = \frac{\text{N} \cdot \text{m}}{\text{coulomb}} = \frac{\text{kg} \cdot \text{m} \cdot \text{m}}{\text{s}^2 \cdot \text{A} \cdot \text{s}} = \frac{\text{kg} \cdot \text{m}^2}{\text{A} \cdot \text{s}^3}.$$

It can also be expressed as amperes times ohms (current times resistance, Ohm's law), watts per ampere (power per unit current, Joule's law), or joules per coulomb (energy per unit charge), which is also equivalent to electron-volts per elementary charge:

$$V = A \cdot \Omega = \frac{W}{A} = \frac{J}{C} = \frac{eV}{e}.$$

Josephson Junction Definition

The "conventional" volt, V_{90}, defined in 1988 by the 18th General Conference on Weights and Measures and in use from 1990, is implemented using the Josephson effect for exact frequency-to-voltage conversion, combined with the caesium frequency standard. For the Josephson constant, $K_J = 2e/h$ (where e is the elementary charge and h is the Planck constant), the "conventional" value K_{J-90} is used:

$$K_{J-90} = 0.4835979 \frac{\text{GHz}}{\mu V}.$$

This standard is typically realized using a series-connected array of several thousand or tens of thousands of junctions, excited by microwave signals between 10 and 80 GHz (depending on the array design). Empirically, several experiments have shown that the method is independent of device design, material, measurement setup, etc., and no correction terms are required in a practical implementation.

Water-flow Analogy

In the *water-flow analogy* sometimes used to explain electric circuits by comparing them with water-filled pipes, voltage (difference in electric potential) is likened to difference in water pressure.

The relationship between voltage and current is defined (in ohmic devices like resistors) by Ohm's Law. Ohm's Law is analogous to the Hagen–Poiseuille equation, as both are linear models relating flux and potential in their respective systems.

Common Voltages

A multimeter can be used to measure the voltage between two positions.

The voltage produced by each electrochemical cell in a battery is determined by the chemistry of that cell. Cells can be combined in series for multiples of that voltage, or additional circuitry added to adjust the voltage to a different level. Mechanical generators can usually be constructed to any voltage in a range of feasibility.

1.5 V C-cell batteries

Nominal voltages of familiar sources:

- Nerve cell resting potential: around −75 mV

- Single-cell, rechargeable NiMH or NiCd battery: 1.2 V

- Single-cell, non-rechargeable (e.g., AAA, AA, C and D cells): alkaline battery: 1.5 V; zinc-carbon battery: 1.56 V if fresh and unused

- LiFePO$_4$ rechargeable battery: 3.3 V

- Lithium polymer rechargeable battery: 3.75 V

- Transistor-transistor logic/CMOS (TTL) power supply: 5 V

- USB: 5 V DC

- PP3 battery: 9 V

- Automobile electrical system: nominal 12 V, about 11.8 V discharged, 12.8 V charged, and 13.8–14.4 V while charging (vehicle running).

- Trucks/lorries: 24 V DC (Europe), 12 V DC (North America)

- Household mains electricity AC:

    - 100 V in Japan

    - 120 V in North America,

    - 230 V in Europe, Asia and Africa,

- Rapid transit third rail: 600–750 V

- High-speed train overhead power lines: 25 kV at 50 Hz

- High-voltage electric power transmission lines: 110 kV and up (1.15 MV was the record as of 2005)

- Lightning: Varies greatly, often around 100 MV.

History

Alessandro Volta

In 1800, as the result of a professional disagreement over the galvanic response advocated by Luigi Galvani, Alessandro Volta developed the so-called voltaic pile, a forerunner of the battery, which produced a steady electric current. Volta had determined that the most effective pair of dissimilar metals to produce electricity was zinc and silver. In the 1880s, the International Electrical Congress, now the International Electrotechnical Commission (IEC), approved the volt as the unit for electromotive force. They made the volt equal to 10^8 cgs units of voltage, the cgs system at the time being the customary system of units in science. They chose such a ratio because the cgs unit of voltage is inconveniently small and one volt in this definition is approximately the emf of a Daniell cell, the standard source of voltage in the telegraph systems of the day. At that time, the volt was defined as the potential difference [i.e., what is nowadays called the "voltage (difference)"] across a conductor when a current of one ampere dissipates one watt of power.

The international volt was defined in 1893 as 1/1.434 of the emf of a Clark cell. This definition was abandoned in 1908 in favor of a definition based on the international ohm and international ampere until the entire set of "reproducible units" was abandoned in 1948.

Prior to the development of the Josephson junction voltage standard, the volt was maintained in national laboratories using specially constructed batteries called standard cells. The United States used a design called the Weston cell from 1905 to 1972.

This SI unit is named after Alessandro Volta. As with every International System of Units (SI) unit named for a person, the first letter of its symbol is upper case (V). However, when an SI unit is spelled out in English, it should always begin with a lower case letter (*volt*)—except in a situation where *any* word in that position would be capitalized, such as at the beginning of a sentence or in material using title case. Note that "degree Celsius" conforms to this rule because the "d" is lowercase.— Based on *The International System of Units*, section 5.2.

Ampere

The ampere (SI unit symbol: A), often shortened to "amp", is the SI unit of electric current (dimension symbol: I) and is one of the seven SI base units. It is named after André-Marie Ampère (1775–1836), French mathematician and physicist, considered the father of electrodynamics.

The ampere is equivalent to one coulomb (roughly 701862410000000000000♠6.241×10¹⁸ times the elementary charge) per second. Amperes are used to express flow rate of electric charge. For any point experiencing a current, if the number of charged particles passing through it — or the charge on the particles passing through it — is increased, the amperes of current at that point will proportionately increase.

The ampere should not be confused with the coulomb (also called "ampere-second") or the ampere-hour (A·h). The ampere is a unit of current, the amount of charge transiting per unit time, and the coulomb is a unit of charge. When SI units are used, constant, instantaneous and average current are expressed in amperes (as in "the charging current is 1.2 A") and the charge accumulated, or passed through a circuit over a period of time is expressed in coulombs (as in "the battery charge is 70043000000000000000♠30000 C"). The relation of the ampere (C/s) to the coulomb is the same as that of the watt (J/s) to the joule.

Definition

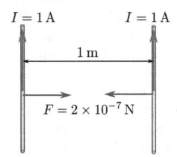

$$I = 1\,\text{A} \qquad\qquad I = 1\,\text{A}$$

$$1\,\text{m}$$

$$F = 2 \times 10^{-7}\,\text{N}$$

Illustration of the definition of the ampere unit

Ampère's force law states that there is an attractive or repulsive force between two parallel wires carrying an electric current. This force is used in the formal definition of the ampere, which states that the ampere is the constant current that will produce an attractive force of 2×10^{-7} newtons per metre of length between two straight, parallel conductors of infinite length and negligible circular cross section placed one metre apart in a vacuum.

The SI unit of charge, the coulomb, "is the quantity of electricity carried in 1 second by a current of 1 ampere". Conversely, a current of one ampere is one coulomb of charge going past a given point per second:

$$1\,\text{A} = 1\,\tfrac{\text{C}}{\text{s}}.$$

In general, charge Q is determined by steady current I flowing for a time t as $Q = It$.

History

The ampere was originally defined as one tenth of the unit of electric current in the centimetre–gram–second system of units; that unit, now known as the abampere, was defined as the amount of current that generates a force of two dynes per centimetre of length between two wires one centimetre apart. The size of the unit was chosen so that the units derived from it in the MKSA system would be conveniently sized.

The "international ampere" was an early realization of the ampere, defined as the current that would deposit 0.001118000 grams of silver per second from a silver nitrate solution. Later, more accurate measurements revealed that this current is 0.99985 A.

Realization

The standard ampere is most accurately realized using a watt balance, but is in practice maintained via Ohm's law from the units of electromotive force and resistance, the volt and the ohm, since the latter two can be tied to physical phenomena that are relatively easy to reproduce, the Josephson junction and the quantum Hall effect, respectively.

At present, techniques to establish the realization of an ampere have a relative uncertainty of approximately a few parts in 10^7, and involve realizations of the watt, the ohm and the volt.

Proposed Future Definition

Rather than a definition in terms of the force between two current-carrying wires, it has been proposed to define the ampere in terms of the rate of flow of elementary charges. Since a coulomb is approximately equal to 6.2415093×10^{18} elementary charges (such as the those carried by protons, or the negative of those carried by electrons), one ampere is approximately equivalent to 6.2415093×10^{18} elementary charges moving past a boundary in one second. (6.2415093×10^{18} is the reciprocal of the value of the elementary charges in coulombs.) The proposed change would define 1 A as being the current in the direction of flow of a particular number of elementary charges per second. In 2005, the International Committee for Weights and Measures (CIPM) agreed to study the proposed change. The new definition was discussed at the 25th General Conference on Weights and Measures (CGPM) in 2014 but for the time being was not adopted.

Everyday Examples

The current drawn by typical constant-voltage energy distribution systems is usually dictated by the power (watt) consumed by the system and the operating voltage. For this reason the examples given below are grouped by voltage level.

Portable Devices

- Hearing aid (typically 1 mW at 1.4 V): 0.7 mA

Internal Combustion Engine Vehicles – 12 V DC

A typical motor vehicle has a 12 V battery. The various accessories that are powered by the battery might include:

- Instrument panel light (typically 2 W): 166 mA.

- Headlights (typically 60 W): 5 A each.

- Starter motor (typically 1–2 kW): 80–160 A

North American Domestic Supply – 120 V AC

Most Canada, Mexico and United States domestic power suppliers run at 120 V.

Household circuit breakers typically provide a maximum of 15 A or 20 A of current to a given set of outlets.

- 22-inch/56-centimeter portable television (35 W): 290 mA
- Tungsten light bulb (60–100 W): 500–830 mA
- Toaster, kettle (1.5 kW): 12.5 A
- Hair dryer (1.8 kW): 15 A

European & Commonwealth Domestic Supply – 230-240 V AC

Most European domestic power supplies run at 230 V, and most Commonwealth domestic power supplies run at 240 V. For the same amount of power (in watts), the current drawn by a particular European or Commonwealth appliance (in Europe or a Commonwealth country) will be less than for an equivalent North American appliance. Typical circuit breakers will provide 16 A.

The current drawn by a number of typical appliances are:

- 22-inch/56-centimeter portable television (35 W): 145–150 mA
- Tungsten light bulb (60–100 W): 240–450 mA
- Compact fluorescent lamp (11–30 W): 56–112 mA
- Toaster, kettle (2 kW): 9 A
- Immersion heater (4.6 kW): 19–20 A

Coulomb

The coulomb (unit symbol: **C**) is the International System of Units (SI) unit of electric charge. It is the charge (symbol: Q or q) transported by a constant current of one ampere in one second:

$$1C = 1A \cdot 1s$$

Thus, it is also the amount of excess charge on a capacitor of one farad charged to a potential difference of one volt:

$$1C = 1F \cdot 1V$$

It is equivalent to the charge of approximately 6.242×10^{18} (1.036×10^{-5} mol) protons, and -1 C is equivalent to the charge of approximately 6.242×10^{18} electrons.

Name and Notation

This SI unit is named after Charles-Augustin de Coulomb. As with every International System of Units (SI) unit named for a person, the first letter of its symbol is upper case (C). However, when an SI unit is spelled out in English, it should always begin with a lower case letter (*coulomb*)—except in a situation where *any* word in that position would be capitalized, such as at the beginning of a sentence or in material using title case. Note that "degree Celsius" conforms to this rule because the "d" is lowercase.— Based on *The International System of Units*, section 5.2.

Definition

The SI system defines the coulomb in terms of the ampere and second: $1\ C = 1\ A \times 1\ s$. The second is defined in terms of a frequency naturally emitted by caesium atoms. The ampere is defined using Ampère's force law; the definition relies in part on the mass of the international prototype kilogram, a metal cylinder housed in France. In practice, the watt balance is used to measure amperes with the highest possible accuracy.

Since the charge of one electron is known to be about $(98\ )\times 10^{-19}$ C, -1 C can also be considered the charge of roughly 6.241509×10^{18} electrons (or $+1$ C the charge of that many positrons or protons), where the number is the reciprocal of 1.602177×10^{-19}.

The proposed redefinition of the ampere and other SI base units would have the effect of fixing the numerical value of the fundamental charge to an explicit constant expressed in coulombs, and therefore it would implicitly fix the value of the coulomb when expressed as a multiple of the fundamental charge (the numerical values of those quantities are the multiplicative inverses of each other).

SI Prefixes

SI multiples for coulomb (C)					
Submultiples			**Multiples**		
Value	**SI symbol**	**Name**	**Value**	**SI symbol**	**Name**
10^{-1} C	dC	decicoulomb	10^{1} C	daC	decacoulomb
10^{-2} C	cC	centicoulomb	10^{2} C	hC	hectocoulomb
10^{-3} C	**mC**	**millicoulomb**	10^{3} C	kC	kilocoulomb
10^{-6} C	**μC**	**microcoulomb**	10^{6} C	MC	megacoulomb
10^{-9} C	**nC**	**nanocoulomb**	10^{9} C	GC	gigacoulomb
10^{-12} C	**pC**	**picocoulomb**	10^{12} C	TC	teracoulomb
10^{-15} C	fC	femtocoulomb	10^{15} C	PC	petacoulomb
10^{-18} C	aC	attocoulomb	10^{18} C	EC	exacoulomb
10^{-21} C	zC	zeptocoulomb	10^{21} C	ZC	zettacoulomb
10^{-24} C	yC	yoctocoulomb	10^{24} C	YC	yottacoulomb
Common multiples are in bold face.					

Conversions

- One coulomb is the magnitude (absolute value) of electrical charge in $6.24150934(14)\times10^{18}$ protons or electrons.

- The inverse of this number gives the elementary charge of $1.6021766208\ (98)\times10^{-19}$ C.

- The magnitude of the electrical charge of one mole of elementary charges (approximately 6.022×10^{23}, or Avogadro's number) is known as a faraday unit of charge (closely related to the Faraday constant). One faraday equals 96485.3399 coulombs. In terms of Avogadro's number (N_A), one coulomb is equal to approximately $1.036 \times N_A \times 10^{-5}$ elementary charges.

- One ampere-hour = 3600 C, 1 mA·h = 3.6 C.

- One statcoulomb (statC), the obsolete CGS electrostatic unit of charge (esu), is approximately 3.3356×10^{-10} C or about one-third of a nanocoulomb.

Relation to Elementary Charge

The elementary charge, the charge of a proton (equivalently, the negative of the charge of an electron), is approximately $1.6021766208(98)\times10^{-19}$ C. In SI, the elementary charge in coulombs is an approximate value: no experiment can be infinitely accurate. However, in other unit systems, the elementary charge has an *exact* value by definition, and other charges are ultimately measured relative to the elementary charge. For example, in conventional electrical units, the values of the Josephson constant K_J and von Klitzing constant R_K are exact defined values (written K_{J-90} and R_{K-90}), and it follows that the elementary charge $e = 2/(K_J R_K)$ is also an exact defined value in this unit system. Specifically, $e_{90} = (2\times10^{-9})/(25812.807 \times 483597.9)$ C exactly. SI itself may someday change its definitions in a similar way. For example, one possible proposed redefinition is "the ampere...is [defined] such that the value of the elementary charge e (charge on a proton) is exactly $1.602176487\times10^{-19}$ coulombs", (in which the numeric value is the 2006 CODATA recommended value, since superseded). This proposal is not yet accepted as part of the SI; the SI definitions are unlikely to change until at least 2015.

In Everyday Terms

- The charges in static electricity from rubbing materials together are typically a few microcoulombs.

- The amount of charge that travels through a lightning bolt is typically around 15 C, although large bolts can be up to 350 C.

- The amount of charge that travels through a typical alkaline AA battery from being fully charged to discharged is about 5 kC = 5000 C ≈ 1400 mA·h.

- According to Coulomb's law, two negative point charges of −1 C, placed one meter apart, would experience a repulsive force of 9×10^9 N, a force roughly equal to the weight of 920000 metric tons of mass on the surface of the Earth.

- The hydraulic analogy uses everyday terms to illustrate movement of charge and the transfer of energy. The analogy equates charge to a volume of water, and voltage to pressure. One coulomb equals (the negative of) the charge of 6.24×10^{18} elec-trons. The amount of energy transferred by the flow of 1 coulomb can vary; for example, 300 times fewer electrons flow through a lightning bolt than in the discharge of an AA bat-tery, but the total energy transferred by the flow of the lightning's electrons is 300 million times greater.

Ohm

The ohm (symbol: Ω) is the SI derived unit of electrical resistance, named after German physicist Georg Simon Ohm. Although several empirically derived standard units for expressing electrical resistance were developed in connection with early telegraphy practice, the British Association for the Advancement of Science proposed a unit derived from existing units of mass, length and time and of a convenient size for practical work as early as 1861. The definition of the ohm was revised several times. Today the definition of the ohm is expressed from the quantum Hall effect.

Definition

A multimeter can be used to measure resistance in ohms, among other things.

The ohm is defined as an electrical resistance between two points of a conductor when a constant potential difference of 1 volt, applied to these points, produces in the conductor a current of 1 ampere, the conductor not being the seat of any electromotive force.

$$\Omega = \frac{V}{A} = \frac{1}{S} = \frac{W}{A^2} = \frac{V^2}{W} = \frac{s}{F} = \frac{J \cdot s}{C^2} = \frac{kg \cdot m^2}{s \cdot C^2} = \frac{J}{s \cdot A^2} = \frac{kg \cdot m^2}{s^3 \cdot A^2}$$

in which the following units appear: volt (V), ampere (A), siemens (S), watt (W), second (s), farad (F), joule (J), kilogram (kg), metre (m), and coulomb (C).

In many cases the resistance of a conductor in ohms is approximately constant within a certain range of voltages, temperatures, and other parameters. These are called linear resistors. In other cases resistance varies (e.g., thermistors).

Commonly used multiples and submultiples in electrical and electronic usage are the microohm, milliohm, kilohm, megohm, and gigaohm; the term "gigohm", though not official, is in common use for the latter.

In alternating current circuits, electrical impedance is also measured in ohms (Ω).

Electrical Engineering: Concepts and Applications

Conversions

The siemens (symbol: S) is the SI derived unit of electric conductance and admittance, also known as the mho (ohm spelled backwards, symbol is ℧); it is the reciprocal of resistance in ohms (Ω).

Power as a Function of Resistance

The power dissipated by a resistor may be calculated from its resistance, and the voltage or current involved. The formula is a combination of Ohm's Law and Joule's law:

$$P = V \cdot I = \frac{V^2}{R} = I^2 \cdot R$$

where:

P = power in watts

R = resistance in ohms

V = voltage across the resistor

I = current through the resistor in amps

A linear resistor has a constant resistance value over all applied voltages or currents; many practical resistors are linear over a useful range of currents. Non-linear resistors have a value that may vary depending on the applied voltage (or current). Where alternating current is applied to the circuit,(or where the resistance value is a function of time), the relation above is true at any instant but calculation of average power over an interval of time will require integration of "instantaneous" power over that interval.

History

The rapid rise of electrotechnology in the last half of the 19th century created a demand for a rational, coherent, consistent, and international system of units for electrical quantities. Telegraphers and other early uses of electricity in the 19th century needed a practical standard unit of measurement for resistance. Resistance was often expressed as a multiple of the resistance of a standard length of telegraph wires; different agencies used different bases for a standard, so units were not readily interchangeable. Electrical units so defined were not a coherent system with the units for energy, mass, length, and time, requiring conversion factors to be used in calculations relating energy or power to resistance.

Two different methods of establishing a system of electrical units can be chosen. Various artifacts, such as a length of wire or a standard electrochemical cell, could be specified as producing defined quantities for resistance, voltage, and so on. Alternatively, the electrical units can be related to the mechanical units by defining, for example, a unit of current that gives a specified force between two wires, or a unit of charge that gives a unit of force between two unit charges. This latter method ensures coherence with the units of energy. Defining a unit for resistance that is coherent with units of energy and time in effect also requires defining units for potential and current. It is desirable that one unit of electrical potential will force one unit of electric current through one unit of electrical resistance, doing one unit of work in one unit of time, otherwise all electrical calculations will require conversion factors.

Since so-called "absolute" units of charge and current are expressed as combinations of units of mass, length, and time, dimensional analysis of the relations between potential, current, and resistance show that resistance is expressed in units of length per time — a velocity. Some early definitions of a unit of resistance, for example, defined a unit resistance as one quadrant of the Earth per second.

The absolute-units system related magnetic and electrostatic quantities to metric base units of mass, time, and length. These units had the great advantage of simplifying the equations used in the solution of electromagnetic problems, and eliminated conversion factors in calculations about electrical quantities. However, the CGS units turned out to have impractical sizes for practical measurements.

Various artifact standards were proposed as the definition of the unit of resistance. In 1860 Werner Siemens published a suggestion for a reproducible resistance standard in Poggendorffs Annalen der Physik und Chemie. He proposed a column of pure mercury, of one square millimetre cross section, one metre long: Siemens mercury unit. However, this unit was not coherent with other units. One proposal was to devise a unit based on a mercury column that would be coherent - in effect, adjusting the length to make the resistance one ohm. Not all users of units had the resources to carry out metrology experiments to the required precision, so working standards notionally based on the physical definition were required.

In 1861, Latimer Clark and Sir Charles Bright presented a paper at the British Association for the Advancement of Science meeting suggesting that standards for electrical units be established and suggesting names for these units derived from eminent philosophers, 'Ohma', 'Farad' and 'Volt'. The BAAS in 1861 appointed a committee including Maxwell and Thomson to report upon Standards of Electrical Resistance. Their objectives were to devise a unit that was of convenient size, part of a complete system for electrical measurements, coherent with the units for energy, stable, reproducible and based on the French metrical system. In the third report of the committee, 1864, the resistance unit is referred to as "B.A. unit, or Ohmad". By 1867 the unit is referred to as simply Ohm.

The B.A. ohm was intended to be 10^9 CGS units but owing to an error in calculations the definition was 1.3% too small. The error was significant for preparation of working standards.

On September 21, 1881 the *Congrès internationale d'électriciens* (international conference of electricians) defined a *practical* unit of Ohm for the resistance, based on CGS units, using a mercury column at zero deg. Celsius, similar to the apparatus suggested by Siemens.

A *legal* ohm, a reproducible standard, was defined by the international conference of electricians at Paris in 1884 as the resistance of a mercury column of specified weight and 106 cm long; this was a compromise value between the B. A. unit (equivalent to 104.7 cm), the Siemens unit (100 cm by definition), and the CGS unit. Although called "legal", this standard was not adopted by any national legislation. The "international" ohm was defined as a mercury column 106.3 cm long of mass 14.4521 grams and 0 °C at the International Electrical Conference 1893 in Chicago. This definition became the basis for the legal definition of the ohm in several countries. In 1908, the next Electrical Conference confirmed this definition. The mercury column standard was maintained until the 1948 General Conference on Weights and Measures, at which the ohm was redefined in absolute terms instead of as an artifact standard.

By the end of the 19th century, units were well-understood and consistent. Definitions would change with little effect on commercial uses of the units. Advances in metrology allowed definitions to be formulated with a high degree of precision and repeatability.

Historical Units of Resistance

Unit	Definition	Value in B.A. ohms	Remarks
Absolute foot/second x 10⁷	using Imperial units	.3048	considered obsolete even in 1884
Thomson's unit	using Imperial units	.3202	100 million feet/second, considered obsolete even in 1884
Jacobi copper unit	A specified copper wire 25 feet long weighing 345 grains	.6367	Used in 1850s
Weber's absolute unit × 10⁷	Based on the metre and the second	0.9191	
Siemens mercury unit	1860. A column of pure mercury	.9537	100 cm and 1 mm² cross section at 0 °C
British Association (B.A.) "ohm"	1863	1.000	Standard coils deposited at Kew Observatory in 1863
Digney, Breguet, Swiss		9.266–10.420	Iron wire 1 km long and 4 square mm cross section
Matthiessen		13.59	One mile of 1/16 inch diameter pure annealed copper wire at 15.5 °C
Varley		25.61	One mile of special 1/16 inch diameter copper wire
German mile		57.44	A German mile (8,238 yard) of iron wire 1/6th inch diameter
Abohm		10^{-9}	Electromagnetic absolute unit in cm-gram-second units
Statohm		$8.987551787 \times 10^{11}$	Electrostatic absolute unit in cm-gram-second units

Realization of Standards

The mercury column method of realizing a physical standard ohm turned out to be difficult to reproduce, owing to the effects of non-constant cross section of the glass tubing. Various resistance coils were constructed by the British Association and others, to serve as physical artifact standards for the unit of resistance. The long-term stability and reproducibility of these artifacts was an ongoing field of research, as the effects of temperature, air pressure, humidity, and time on the standards were detected and analyzed.

Artifact standards are still used, but metrology experiments relating accurately-dimensioned inductors and capacitors provided a more fundamental basis for the definition of the ohm. Since

1990 the quantum Hall effect has been used to define the ohm with high precision and repeatability. The quantum Hall experiments are used to check the stability of working standards that have convenient values for comparison.

Symbol

When preparing electronic documents, some document editing software applications have used the Symbol typeface to render the character Ω. Where the font is not supported, a W is displayed instead ("10 W" instead of "10 Ω", for instance). As W represents the watt, the SI unit of power, this can lead to confusion.

In the electronics industry it is common to use the character R instead of the Ω symbol, thus, a 10 Ω resistor may be represented as 10R. This is the British standard BS 1852 code. It is used in many instances where the value has a decimal place. For example, 5.6 Ω is listed as 5R6. This method avoids overlooking the decimal period, which may not be rendered reliably on components or when duplicating documents.

In documents printed before WWII the unit symbol often consisted of the raised lowercase omega (ω), such that 56 Ω was written as 56$^{\omega}$.

Unicode encodes the symbol as U+2126 Ω OHM SIGN, distinct from Greek omega among letter-like symbols, but it is only included for backwards compatibility and the Greek uppercase omega character U+03A9 Ω GREEK CAPITAL LETTER OMEGA (HTML Ω • Ω) is preferred. In DOS and Windows, the alt code ALT 234 may produce the Ω symbol. In Mac OS, Opt+Z does the same.

Farad

The farad (symbol: F) is the SI derived unit of electrical capacitance, the ability of a body to store an electrical charge. It is named after the English physicist Michael Faraday.

Definition

One farad is defined as the capacitance across which, when charged with one coulomb, there is a potential difference of one volt. Equally, one farad can be described as the capacitance which stores a one coulomb charge across a potential difference of one volt.

The relationship between capacitance, charge and potential difference is linear. For example, if the potential difference across a capacitor is halved, the quantity of charge stored by that capacitor will also be halved.

For most applications, the farad is an impractically large unit of capacitance. Most electrical and electronic applications are covered by the following SI prefixes:

- 1 mF (millifarad, one thousandth (10^{-3}) of a farad) = 1000 μF = 1000000 nF

- 1 μF (microfarad, one millionth (10^{-6}) of a farad) = 1000 nF = 1000000 pF

- 1 nF (nanofarad, one billionth (10^{-9}) of a farad) = 1000 pF
- 1 pF (picofarad, one trillionth (10^{-12}) of a farad)

Equalities

A farad has the base SI representation of: $s^4 \times A^2 \times m^{-2} \times kg^{-1}$

It can further be expressed as:

$$F = \frac{A \cdot s}{V} = \frac{J}{V^2} = \frac{W \cdot s}{V^2} = \frac{C}{V} = \frac{C^2}{J} = \frac{C^2}{N \cdot m} = \frac{s^2 \cdot C^2}{m^2 \cdot kg} = \frac{s^4 \cdot A^2}{m^2 \cdot kg} = \frac{s}{\Omega} = \frac{s^2}{H}$$

where F=farad, A=ampere, V=volt, C=coulomb, J=joule, m=metre, N=newton, s=second, W=watt, kg=kilogram, Ω=ohm, H=henry.

History

The term "farad" was originally coined by Latimer Clark and Charles Bright in 1861, in honor of Michael Faraday, for a unit of quantity of charge but, starting in 1881 at the International Congress of Electricians in Paris, the name farad was officially used for the unit of electrical capacitance.

Explanation

Examples of different types of capacitors

A capacitor consists of two conducting surfaces, frequently referred to as plates, separated by an insulating layer usually referred to as a dielectric. The original capacitor was the Leyden jar developed in the 18th century. It is the accumulation of electric charge on the plates that results in capacitance. Modern capacitors are constructed using a range of manufacturing techniques and materials to provide the extraordinarily wide range of capacitance values used in electronics applications from femtofarads to farads, with maximum-voltage ratings ranging from a few volts to several kilovolts.

Values of capacitors are usually specified in farads (F), microfarads (μF), nanofarads (nF) and picofarads (pF). The millifarad is rarely used in practice (a capacitance of 4.7 mF (0.0047 F), for example, is instead written as 4700 μF), while the nanofarad is uncommon in North America. The size of commercially available capacitors ranges from around 0.1 pF to 5000F (5 kF) supercapacitors. Parasitic capacitance in high-performance integrated circuits can be measured in femtofarads (1 fF = 0.001 pF = 10^{-15} F), while high-performance test equipment can detect changes in capacitance on the order of tens of attofarads (1 aF = 10^{-18} F).

A value of 0.1 pF is about the smallest available in capacitors for general use in electronic design, since smaller ones would be dominated by the parasitic capacitances of other components, wiring or printed circuit boards. Capacitance values of 1 pF or lower can be achieved by twisting two short lengths of insulated wire together.

The capacitance of the Earth's ionosphere with respect to the ground is calculated to be about 1 F.

Informal and Deprecated Terminology

The picofarad is sometimes colloquially pronounced as "puff" or "pic", as in "a ten-puff capacitor". Similarly, "mic" (pronounced "mike") is sometimes used informally to signify microfarads. If the Greek letter μ is not available, the notation "uF" is often used as a substitute for "μF" in electronics literature. A "micro-microfarad" (μμF, and confusingly often mmf or MMF), an obsolete unit sometimes found in older texts, is the equivalent of a picofarad. In texts prior to 1960, and on capacitor packages even much more recently, mf or MFD rather than the modern μF frequently represented microfarads. Similarly, mmf represented picofarads.

Related Concepts

The reciprocal of capacitance is called electrical elastance, the (non-standard, non-SI) unit of which is the daraf.

CGS Units

The abfarad (abbreviated abF) is an obsolete CGS unit of capacitance equal to 10^9 farads (1 gigafarad, GF).

The statfarad (abbreviated statF) is a rarely used CGS unit equivalent to the capacitance of a capacitor with a charge of 1 statcoulomb across a potential difference of 1 statvolt. It is $1/(10^{-5}c^2)$ farad, approximately 1.1126 picofarads.

Henry (Unit)

The henry (symbol H) is the unit of electrical inductance in the International System of Units. The unit is named after Joseph Henry (1797–1878), the American scientist who discovered electromagnetic induction independently of and at about the same time as Michael Faraday (1791–1867) in England. The magnetic permeability of vacuum is $4\pi \times 10^{-7}$ H m^{-1} (henry per metre).

The United States National Institute of Standards and Technology recommends English-speaking users of SI to write the plural as *henries*.

Definition

The inductance of an electric circuit is one henry when an electric current that is changing at one ampere per second results in an electromotive force of one volt across the inductor :

$$v(t) = L\frac{di}{dt},$$

where *v(t)* denotes the resulting voltage across the circuit, *i(t)* is the current through the circuit, and *L* is the inductance of the circuit.

The henry is a derived unit based on four of the seven base units of the International System of Units: kilogram (kg), meter (m), second (s), and ampere (A). Expressed in combinations of SI units, the henry is:

$$H = \frac{kg \cdot m^2}{s^2 \cdot A^2} = \frac{kg \cdot m^2}{C^2} = \frac{J}{A^2} = \frac{T \cdot m^2}{A} = \frac{Wb}{A} = \frac{V \cdot s}{A} = \frac{s^2}{F} = \frac{1}{F \cdot Hz^2} = \Omega \cdot s$$

where

Wb = weber,
T = tesla,
J = joule,
m = meter,
s = second,
A = ampere,
V = volt,
C = coulomb,
F = farad,
Hz = hertz,
Ω = ohm

As with every International System of Units (SI) unit named for a person, the first letter of its symbol is upper case (H). However, when an SI unit is spelled out in English, it should always begin with a lower case letter (henry)—except in a situation where any word in that position would be capitalized, such as at the beginning of a sentence or in material using title case. Note that "degree Celsius" conforms to this rule because the "d" is lowercase.— Based on The International System of Units , section 5.2.

References

- Hill, Paul Horowitz; Winfield; Winfield, Hill (2015). The Art of Electronics (3. ed.). Cambridge [u.a.]: Cambridge Univ. Press. p. 689. ISBN 978-0-521-809269.

- Serway, Raymond A; Jewett, JW (2006). Serway's principles of physics: a calculus based text (Fourth ed.). Belmont, CA: Thompson Brooks/Cole. p. 746. ISBN 0-53449143-X.

- Monk, Paul MS (2004), Physical Chemistry: Understanding our Chemical World, John Wiley & Sons, ISBN 0-471-49180-2 .

- Tunbridge, Paul (1992). Lord Kelvin : his influence on electrical measurements and units. London: Peregrinus. pp. 26, 39–40. ISBN 9780863412370. Retrieved 5 May 2015.

- Platt, Charles (2009). Make: Electronics: Learning Through Discovery. O'Reilly Media. p. 61. ISBN 9781449388799. Retrieved 2014-07-22. Nanofarads are also used, more often in Europe than in the United States.

- "CODATA Value: elementary charge". The NIST Reference on Constants, Units, and Uncertainty. US National Institute of Standards and Technology. June 2015. Retrieved 2015-09-22.

- Hunt, Bruce J (1994). "The Ohm Is Where the Art Is: British Telegraph Engineers and the Development of Electrical Standards" (PDF). Osiris. 2nd. 9: 48–63. doi:10.1086/368729. Retrieved 27 February 2014.

- Clark, Latimer; Bright, Sir Charles (1861-11-09). "Measurement of Electrical Quantities and Resistance". The Electrician. 1 (1): 3–4. Retrieved 27 February 2014.

Applications of Electrical Engineering

A number of applications are based on electrical engineering. Some of these are avionics, transistor, voltmeter, electrical telegraph and electric motor. Avionics are used in satellites and spacecrafts and it helps in the navigation and management of the systems found in these vehicles. The aspects elucidated in this chapter are of vital importance, and provide a better understanding of electrical engineering.

Avionics

Avionics are the electronic systems used on aircraft, artificial satellites, and spacecraft. Avionic systems include communications, navigation, the display and management of multiple systems, and the hundreds of systems that are fitted to aircraft to perform individual functions. These can be as simple as a searchlight for a police helicopter or as complicated as the tactical system for an airborne early warning platform. The term *avionics* is a portmanteau of the words *aviation* and *electronics*.

Radar and other avionics in the nose of a Cessna Citation I/SP

History

The term avionics was coined by the journalist Philip J. Klass as a portmanteau of aviation electronics. Many modern avionics have their origins in World War II wartime developments. For example, autopilot systems that are prolific today were started to help bomber planes fly steadily enough to hit precision targets from high altitudes. Famously, radar was developed in the UK, Germany, and the United States during the same period. Modern avionics is a substantial portion of military aircraft spending. Aircraft like the F15E and the now retired F14 have roughly 20 percent of their budget spent on avionics. Most modern helicopters now have budget splits of 60/40 in favour of avionics.

The civilian market has also seen a growth in cost of avionics. Flight control systems (fly-by-wire) and new navigation needs brought on by tighter airspaces, have pushed up development costs. The major change has been the recent boom in consumer flying. As more people begin to use planes as their primary method of transportation, more elaborate methods of controlling aircraft safely in these high restrictive airspaces have been invented.

Modern Avionics

Avionics plays a heavy role in modernization initiatives like the Federal Aviation Administration's (FAA) Next Generation Air Transportation System project in the United States and the Single European Sky ATM Research (SESAR) initiative in Europe. The Joint Planning and Development Office put forth a roadmap for avionics in six areas:

- Published Routes and Procedures – Improved navigation and routing

- Negotiated Trajectories – Adding data communications to create preferred routes dynamically

- Delegated Separation – Enhanced situational awareness in the air and on the ground

- LowVisibility/CeilingApproach/Departure – Allowing operations with weather constraints with less ground infrastructure

- Surface Operations – To increase safety in approach and departure

- ATM Efficiencies – Improving the ATM process

Aircraft Avionics

The cockpit of an aircraft is a typical location for avionic equipment, including control, monitoring, communication, navigation, weather, and anti-collision systems. The majority of aircraft power their avionics using 14- or 28volt DC electrical systems; however, larger, more sophisticated aircraft (such as airliners or military combat aircraft) have AC systems operating at 400 Hz, 115 volts AC. There are several major vendors of flight avionics, including Panasonic Avionics Corporation, Honeywell (which now owns Bendix/King), Rockwell Collins, Thales Group, GE Aviation Systems, Garmin, Parker Hannifin, UTC Aerospace Systems and Avidyne Corporation.

One source of international standards for avionics equipment are prepared by the Airlines Electronic Engineering Committee (AEEC) and published by ARINC.

Communications

Communications connect the flight deck to the ground and the flight deck to the passengers. Onboard communications are provided by public-address systems and aircraft intercoms.

The VHF aviation communication system works on the airband of 118.000 MHz to 136.975 MHz. Each channel is spaced from the adjacent ones by 8.33 kHz in Europe, 25 kHz elsewhere. VHF is also used for line of sight communication such as aircraft-to-aircraft and aircraft-to-ATC. Amplitude modulation (AM) is used, and the conversation is performed in simplex mode. Aircraft communication can also take place using HF (especially for trans-oceanic flights) or satellite communication.

Navigation

Navigation is the determination of position and direction on or above the surface of the Earth. Avionics can use satellite-based systems (such as GPS and WAAS), ground-based systems (such as VOR or LORAN), or any combination thereof. Navigation systems calculate the position automatically and display it to the flight crew on moving map displays. Older avionics required a pilot or navigator to plot the intersection of signals on a paper map to determine an aircraft's location; modern systems calculate the position automatically and display it to the flight crew on moving map displays.

Monitoring

The Airbus A380 glass cockpit featuring pull-out keyboards and two wide computer screens on the sides for pilots.

The first hints of glass cockpits emerged in the 1970s when flight-worthy cathode ray tubes (CRT) screens began to replace electromechanical displays, gauges and instruments. A "glass" cockpit refers to the use of computer monitors instead of gauges and other analog displays. Aircraft were getting progressively more displays, dials and information dashboards that eventually competed for space and pilot attention. In the 1970s, the average aircraft had more than 100 cockpit instruments and controls.

Glass cockpits started to come into being with the Gulfstream GIV private jet in 1985. One of the key challenges in glass cockpits is to balance how much control is automated and how much the pilot should do manually. Generally they try to automate flight operations while keeping the pilot constantly informed.

Aircraft Flight-control Systems

Aircraft have means of automatically controlling flight. Autopilot was first invented by Lawrence Sperry during World War I to fly bomber planes steady enough to hit precision targets from 25,000 feet. When it was first adopted by the U.S. military, a Honeywell engineer sat in the back seat with bolt cutters to disconnect the autopilot in case of emergency. Nowadays most commercial planes are equipped with aircraft flight control systems in order to reduce pilot error and workload at landing or takeoff.

The first simple commercial auto-pilots were used to control heading and altitude and had limited authority on things like thrust and flight control surfaces. In helicopters, auto-stabilization was used in a similar way. The first systems were electromechanical. The advent of fly by wire and electro-actuated flight surfaces (rather than the traditional hydraulic) has increased safety. As with displays and instruments, critical devices that were electro-mechanical had a finite life. With safety critical systems, the software is very strictly tested.

Collision-avoidance Systems

To supplement air traffic control, most large transport aircraft and many smaller ones use a traffic alert and collision avoidance system (TCAS), which can detect the location of nearby aircraft, and provide instructions for avoiding a midair collision. Smaller aircraft may use simpler traffic alerting systems such as TPAS, which are passive (they do not actively interrogate the transponders of other aircraft) and do not provide advisories for conflict resolution.

To help avoid controlled flight into terrain (CFIT), aircraft use systems such as ground-proximity warning systems (GPWS), which use radar altimeters as a key element. One of the major weaknesses of GPWS is the lack of "look-ahead" information, because it only provides altitude above terrain "look-down". In order to overcome this weakness, modern aircraft use a terrain awareness warning system (TAWS).

Black Boxes

Commercial aircraft cockpit data recorders, commonly known as a "black box", store flight information and audio from the cockpit. They are often recovered from a plane after a crash to determine control settings and other parameters during the incident.

Weather Systems

Weather systems such as weather radar (typically Arinc 708 on commercial aircraft) and lightning detectors are important for aircraft flying at night or in instrument meteorological conditions, where it is not possible for pilots to see the weather ahead. Heavy precipitation (as sensed by radar) or severe turbulence (as sensed by lightning activity) are both indications of strong convective activity and severe turbulence, and weather systems allow pilots to deviate around these areas.

Lightning detectors like the Stormscope or Strikefinder have become inexpensive enough that they are practical for light aircraft. In addition to radar and lightning detection, observations and extended radar pictures (such as NEXRAD) are now available through satellite data connections, allowing pilots to see weather conditions far beyond the range of their own in-flight systems. Modern displays allow weather information to be integrated with moving maps, terrain, and traffic onto a single screen, greatly simplifying navigation.

Modern weather systems also include wind shear and turbulence detection and terrain and traffic warning systems. Inplane weather avionics are especially popular in Africa, India, and other countries where air-travel is a growing market, but ground support is not as well developed.

Aircraft Management Systems

There has been a progression towards centralized control of the multiple complex systems fitted to aircraft, including engine monitoring and management. Health and usage monitoring systems

(HUMS) are integrated with aircraft management computers to give maintainers early warnings of parts that will need replacement.

The integrated modular avionics concept proposes an integrated architecture with application software portable across an assembly of common hardware modules. It has been used in fourth generation jet fighters and the latest generation of airliners.

Mission or Tactical Avionics

Military aircraft have been designed either to deliver a weapon or to be the eyes and ears of other weapon systems. The vast array of sensors available to the military is used for whatever tactical means required. As with aircraft management, the bigger sensor platforms (like the E3D, JSTARS, ASTOR, Nimrod MRA4, Merlin HM Mk 1) have mission-management computers.

Police and EMS aircraft also carry sophisticated tactical sensors.

Military Communications

While aircraft communications provide the backbone for safe flight, the tactical systems are designed to withstand the rigors of the battle field. UHF, VHF Tactical (30–88 MHz) and SatCom systems combined with ECCM methods, and cryptography secure the communications. Data links such as Link 11, 16, 22 and BOWMAN, JTRS and even TETRA provide the means of transmitting data (such as images, targeting information etc.).

Radar

Airborne radar was one of the first tactical sensors. The benefit of altitude providing range has meant a significant focus on airborne radar technologies. Radars include airborne early warning (AEW), anti-submarine warfare (ASW), and even weather radar (Arinc 708) and ground tracking/proximity radar.

The military uses radar in fast jets to help pilots fly at low levels. While the civil market has had weather radar for a while, there are strict rules about using it to navigate the aircraft.

Sonar

Dipping sonar fitted to a range of military helicopters allows the helicopter to protect shipping assets from submarines or surface threats. Maritime support aircraft can drop active and passive sonar devices (sonobuoys) and these are also used to determine the location of hostile submarines.

Electro-optics

Electro-optic systems include devices such as the head-up display (HUD), forward looking infra-red (FLIR), infra-red search and track and other passive infrared devices (Passive infrared sensor). These are all used to provide imagery and information to the flight crew. This imagery is used for everything from search and rescue to navigational aids and target acquisition.

ESM/DAS

Electronic support measures and defensive aids are used extensively to gather information about threats or possible threats. They can be used to launch devices (in some cases automatically) to counter direct threats against the aircraft. They are also used to determine the state of a threat and identify it.

Aircraft Networks

The avionics systems in military, commercial and advanced models of civilian aircraft are interconnected using an avionics databus. Common avionics databus protocols, with their primary application, include:

- Aircraft Data Network (ADN): Ethernet derivative for Commercial Aircraft
- Avionics Full-Duplex Switched Ethernet (AFDX): Specific implementation of ARINC 664 (ADN) for Commercial Aircraft
- ARINC 429: Generic Medium-Speed Data Sharing for Private and Commercial Aircraft
- ARINC 664: ADN
- ARINC 629: Commercial Aircraft (Boeing 777)
- ARINC 708: Weather Radar for Commercial Aircraft
- ARINC 717: Flight Data Recorder for Commercial Aircraft
- ARINC 825: CAN bus for commercial aircraft (for example Boeing 787 and Airbus A350)
- IEEE 1394b: Military Aircraft
- MIL-STD-1553: Military Aircraft
- MIL-STD-1760: Military Aircraft
- TTP – Time-Triggered Protocol: Boeing 787 Dreamliner, Airbus A380, Fly-By-Wire Actuation Platforms from Parker Aerospace
- TTEthernet – Time-Triggered Ethernet: NASA Orion Spacecraft

Transistor

A transistor is a semiconductor device used to amplify or switch electronic signals and electrical power. It is composed of semiconductor material usually with at least three terminals for connection to an external circuit. A voltage or current applied to one pair of the transistor's terminals changes the current through another pair of terminals. Because the controlled (output) power can be higher than the controlling (input) power, a transistor can amplify a signal. Today, some transistors are packaged individually, but many more are found embedded in integrated circuits.

Assorted discrete transistors. Packages in order from top to bottom: TO-3, TO-126, TO-92, SOT-23.

The transistor is the fundamental building block of modern electronic devices, and is ubiquitous in modern electronic systems. First conceived by Julius Lilienfeld in 1926 and practically implemented in 1947 by American physicists John Bardeen, Walter Brattain, and William Shockley, the transistor revolutionized the field of electronics, and paved the way for smaller and cheaper radios, calculators, and computers, among other things. The transistor is on the list of IEEE milestones in electronics, and Bardeen, Brattain, and Shockley shared the 1956 Nobel Prize in Physics for their achievement.

History

A replica of the first working transistor.

The thermionic triode, a vacuum tube invented in 1907, enabled amplified radio technology and long-distance telephony. The triode, however, was a fragile device that consumed a lot of power. Physicist Julius Edgar Lilienfeld filed a patent for a field-effect transistor (FET) in Canada in 1925, which was intended to be a solid-state replacement for the triode. Lilienfeld also filed identical patents in the United States in 1926 and 1928. However, Lilienfeld did not publish any research articles about his devices nor did his patents cite any specific examples of a working prototype. Because the production of high-quality semiconductor materials was still decades away, Lilienfeld's solid-state amplifier ideas would not have found practical use in the 1920s and 1930s, even if such a device had been built. In 1934, German inventor Oskar Heil patented a similar device in Europe.

John Bardeen, William Shockley and Walter Brattain at Bell Labs, 1948.

From November 17, 1947 to December 23, 1947, John Bardeen and Walter Brattain at AT&T's Bell Labs in the United States performed experiments and observed that when two gold point contacts were applied to a crystal of germanium, a signal was produced with the output power greater than the input. Solid State Physics Group leader William Shockley saw the potential in this, and over the next few months worked to greatly expand the knowledge of semiconductors. The term *transistor* was coined by John R. Pierce as a contraction of the term *transresistance*. According to Lillian Hoddeson and Vicki Daitch, authors of a biography of John Bardeen, Shockley had proposed that Bell Labs' first patent for a transistor should be based on the field-effect and that he be named as the inventor. Having unearthed Lilienfeld's patents that went into obscurity years earlier, lawyers at Bell Labs advised against Shockley's proposal because the idea of a field-effect transistor that used an electric field as a "grid" was not new. Instead, what Bardeen, Brattain, and Shockley invented in 1947 was the first point-contact transistor. In acknowledgement of this accomplishment, Shockley, Bardeen, and Brattain were jointly awarded the 1956 Nobel Prize in Physics "for their researches on semiconductors and their discovery of the transistor effect."

Herbert F. Mataré (1950)

In 1948, the point-contact transistor was independently invented by German physicists Herbert Mataré and Heinrich Welker while working at the Compagnie des Freins et Signaux, a Westinghouse subsidiary located in Paris. Mataré had previous experience in developing crystal rectifiers from silicon and germanium in the German radar effort during World War II. Using this knowledge, he began researching the phenomenon of "interference" in 1947. By June 1948, witnessing currents flowing through point-contacts, Mataré produced consistent results using samples of germanium produced by Welker, similar to what Bardeen and Brattain had accomplished earlier in December 1947. Realizing that Bell Labs' scientists had already invented the transistor before them, the company rushed to get its "transistron" into production for amplified use in France's telephone network.

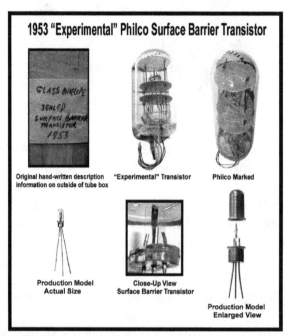

Philco surface-barrier transistor developed and produced in 1953

The first high-frequency transistor was the surface-barrier germanium transistor developed by Philco in 1953, capable of operating up to 60 MHz. These were made by etching depressions into an N-type germanium base from both sides with jets of Indium(III) sulfate until it was a few ten-thousandths of an inch thick. Indium electroplated into the depressions formed the collector and emitter.

The first "prototype" pocket transistor radio was shown by INTERMETALL (a company founded by Herbert Mataré in 1952) at the Internationale Funkausstellung Düsseldorf between August 29, 1953 and September 9, 1953.

The first "production" all-transistor car radio was produced in 1955 by Chrysler and Philco, had used surface-barrier transistors in its circuitry and which were also first suitable for high-speed computers.

The first working silicon transistor was developed at Bell Labs on January 26, 1954 by Morris Tanenbaum. The first commercial silicon transistor was produced by Texas Instruments in 1954. This was the work of Gordon Teal, an expert in growing crystals of high purity, who had previously worked at Bell Labs. The first MOS transistor actually built was by Kahng and Atalla at Bell Labs in 1960.

Importance

A Darlington transistor opened up so the actual transistor chip (the small square) can be seen inside. A Darlington transistor is effectively two transistors on the same chip. One transistor is much larger than the other, but both are large in comparison to transistors in large-scale integration because this particular example is intended for power applications.

The transistor is the key active component in practically all modern electronics. Many consider it to be one of the greatest inventions of the 20th century. Its importance in today's society rests on its ability to be mass-produced using a highly automated process (semiconductor device fabrication) that achieves astonishingly low per-transistor costs. The invention of the first transistor at Bell Labs was named an IEEE Milestone in 2009.

Although several companies each produce over a billion individually packaged (known as *discrete*) transistors every year, the vast majority of transistors are now produced in integrated circuits (often shortened to *IC*, *microchips* or simply *chips*), along with diodes, resistors, capacitors and other electronic components, to produce complete electronic circuits. A logic gate consists of up to about twenty transistors whereas an advanced microprocessor, as of 2009, can use as many as 3 billion transistors (MOSFETs). "About 60 million transistors were built in 2002... for [each] man, woman, and child on Earth."

The transistor's low cost, flexibility, and reliability have made it a ubiquitous device. Transistorized mechatronic circuits have replaced electromechanical devices in controlling appliances and machinery. It is often easier and cheaper to use a standard microcontroller and write a computer program to carry out a control function than to design an equivalent mechanical control function.

Simplified Operation

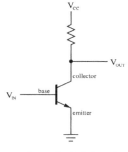

A simple circuit diagram to show the labels of a n–p–n bipolar transistor.

The essential usefulness of a transistor comes from its ability to use a small signal applied between one pair of its terminals to control a much larger signal at another pair of terminals. This property is called gain. It can produce a stronger output signal, a voltage or current, which is proportional to a weaker input signal; that is, it can act as an amplifier. Alternatively, the transistor can be used to turn current on or off in a circuit as an electrically controlled switch, where the amount of current is determined by other circuit elements.

There are two types of transistors, which have slight differences in how they are used in a circuit. A *bipolar transistor* has terminals labeled base, collector, and emitter. A small current at the base terminal (that is, flowing between the base and the emitter) can control or switch a much larger current between the collector and emitter terminals. For a *field-effect transistor*, the terminals are labeled gate, source, and drain, and a voltage at the gate can control a current between source and drain.

The image represents a typical bipolar transistor in a circuit. Charge will flow between emitter and collector terminals depending on the current in the base. Because internally the base and emitter connections behave like a semiconductor diode, a voltage drop develops between base and emitter while the base current exists. The amount of this voltage depends on the material the transistor is made from, and is referred to as V_{BE}.

Transistor as a Switch

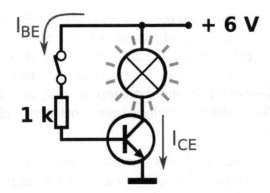

BJT used as an electronic switch, in grounded-emitter configuration.

Transistors are commonly used in digital circuits as electronic switches which can be either in an "on" or "off" state, both for high-power applications such as switched-mode power supplies and for low-power applications such as logic gates. Important parameters for this application include the current switched, the voltage handled, and the switching speed, characterised by the rise and fall times.

In a grounded-emitter transistor circuit, such as the light-switch circuit shown, as the base voltage rises, the emitter and collector currents rise exponentially. The collector voltage drops because of reduced resistance from collector to emitter. If the voltage difference between the collector and emitter were zero (or near zero), the collector current would be limited only by the load resistance (light bulb) and the supply voltage. This is called *saturation* because current is flowing from collector to emitter freely. When saturated, the switch is said to be *on*.

Providing sufficient base drive current is a key problem in the use of bipolar transistors as switches. The transistor provides current gain, allowing a relatively large current in the collector to be switched by a much smaller current into the base terminal. The ratio of these currents varies depending on the type of transistor, and even for a particular type, varies depending on the collector current. In the example light-switch circuit shown, the resistor is chosen to provide enough base current to ensure the transistor will be saturated.

In a switching circuit, the idea is to simulate, as near as possible, the ideal switch having the properties of open circuit when off, short circuit when on, and an instantaneous transition between the two states. Parameters are chosen such that the "off" output is limited to leakage currents too small to affect connected circuitry; the resistance of the transistor in the "on" state is too small to affect circuitry; and the transition between the two states is fast enough not to have a detrimental effect.

Transistor as an Amplifier

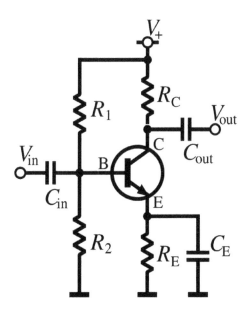

Amplifier circuit, common-emitter configuration with a voltage-divider bias circuit.

The common-emitter amplifier is designed so that a small change in voltage (V_{in}) changes the small current through the base of the transistor; the transistor's current amplification combined with the properties of the circuit mean that small swings in V_{in} produce large changes in V_{out}.

Various configurations of single transistor amplifier are possible, with some providing current gain, some voltage gain, and some both.

From mobile phones to televisions, vast numbers of products include amplifiers for sound reproduction, radio transmission, and signal processing. The first discrete-transistor audio amplifiers barely supplied a few hundred milliwatts, but power and audio fidelity gradually increased as better transistors became available and amplifier architecture evolved.

Modern transistor audio amplifiers of up to a few hundred watts are common and relatively inexpensive.

Comparison with Vacuum Tubes

Before transistors were developed, vacuum (electron) tubes (or in the UK "thermionic valves" or just "valves") were the main active components in electronic equipment.

Advantages

The key advantages that have allowed transistors to replace vacuum tubes in most applications are

- no cathode heater (which produces the characteristic orange glow of tubes), reducing power consumption, eliminating delay as tube heaters warm up, and immune from cathode poisoning and depletion;

- very small size and weight, reducing equipment size;

- large numbers of extremely small transistors can be manufactured as a single integrated circuit;

- low operating voltages compatible with batteries of only a few cells;

- circuits with greater energy efficiency are usually possible. For low-power applications (e.g., voltage amplification) in particular, energy consumption can be very much less than for tubes;

- inherent reliability and very long life; tubes always degrade and fail over time. Some transistorized devices have been in service for more than 50 years ;

- complementary devices available, providing design flexibility including complementary-symmetry circuits, not possible with vacuum tubes;

- very low sensitivity to mechanical shock and vibration, providing physical ruggedness and virtually eliminating shock-induced spurious signals (e.g., microphonics in audio applications);

- not susceptible to breakage of a glass envelope, leakage, outgassing, and other physical damage.

Limitations

Transistors have the following limitations:

- silicon transistors can age and fail;

- high-power, high-frequency operation, such as that used in over-the-air television broadcasting, is better achieved in vacuum tubes due to improved electron mobility in a vacuum;

- solid-state devices are susceptible to damage from very brief electrical and thermal events, including electrostatic discharge in handling; vacuum tubes are electrically much more rugged;

- sensitivity to radiation and cosmic rays (special radiation-hardened chips are used for spacecraft devices);

- vacuum tubes in audio applications create significant lower-harmonic distortion, the so-called tube sound, which some people prefer.

Types

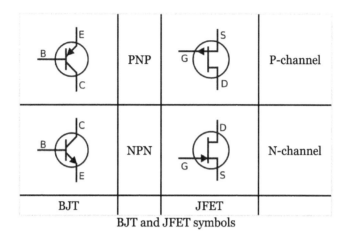

BJT and JFET symbols

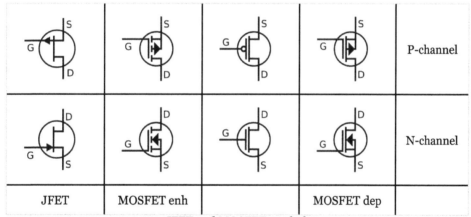

JFET and MOSFET symbols

Transistors are categorized by

- semiconductor material: the metalloids germanium (first used in 1947) and silicon (first used in 1954)—in amorphous, polycrystalline and monocrystalline form—, the compounds gallium arsenide (1966) and silicon carbide (1997), the alloy silicon-germanium (1989), the allotrope of carbon graphene (research ongoing since 2004), etc.

- structure: BJT, JFET, IGFET (MOSFET), insulated-gate bipolar transistor, "other types";

- electrical polarity (positive and negative): n–p–n, p–n–p (BJTs), n-channel, p-channel (FETs);

- maximum power rating: low, medium, high;

- maximum operating frequency: low, medium, high, radio (RF), microwave frequency (the maximum effective frequency of a transistor in a common-emitter or common-source circuit is denoted by the term f_T, an abbreviation for transition frequency—the frequency of transition is the frequency at which the transistor yields unity voltage gain)

- application: switch, general purpose, audio, high voltage, super-beta, matched pair;

- physical packaging: through-hole metal, through-hole plastic, surface mount, ball grid array, power modules.

- amplification factor h_{FE}, β_F (transistor beta) or g_m (transconductance).

Hence, a particular transistor may be described as *silicon, surface-mount, BJT, n–p–n, low-power, high-frequency switch.*

A popular way to remember which symbol represents which type of transistor is to look at the arrow and how it is arranged. Within an NPN transistor symbol, the arrow will Not Point iN. Conversely, within the PNP symbol you see that the arrow Points iN Proudly.

Bipolar Junction Transistor (BJT)

Bipolar transistors are so named because they conduct by using both majority and minority carriers. The bipolar junction transistor, the first type of transistor to be mass-produced, is a combination of two junction diodes, and is formed of either a thin layer of p-type semiconductor sandwiched between two n-type semiconductors (an n–p–n transistor), or a thin layer of n-type semiconductor sandwiched between two p-type semiconductors (a p–n–p transistor). This construction produces two p–n junctions: a base–emitter junction and a base–collector junction, separated by a thin region of semiconductor known as the base region (two junction diodes wired together without sharing an intervening semiconducting region will not make a transistor).

BJTs have three terminals, corresponding to the three layers of semiconductor—an *emitter*, a *base*, and a *collector*. They are useful in amplifiers because the currents at the emitter and collector are controllable by a relatively small base current. In an n–p–n transistor operating in the active region, the emitter–base junction is forward biased (electrons and holes recombine at the junction), and electrons are injected into the base region. Because the base is narrow, most of these electrons will diffuse into the reverse-biased (electrons and holes are formed at, and move away from the junction) base–collector junction and be swept into the collector; perhaps one-hundredth of the electrons will recombine in the base, which is the dominant mechanism in the base current. By controlling the number of electrons that can leave the base, the number of electrons entering the collector can be controlled. Collector current is approximately β (common-emitter current gain) times the base current. It is typically greater than 100 for small-signal transistors but can be smaller in transistors designed for high-power applications.

Unlike the field-effect transistor, the BJT is a low-input-impedance device. Also, as the base–emitter voltage (V_{BE}) is increased the base–emitter current and hence the collector–emitter current (I_{CE}) increase exponentially according to the Shockley diode model and the Ebers-Moll model. Because of this exponential relationship, the BJT has a higher transconductance than the FET.

Bipolar transistors can be made to conduct by exposure to light, because absorption of photons in the base region generates a photocurrent that acts as a base current; the collector current is approximately β times the photocurrent. Devices designed for this purpose have a transparent window in the package and are called phototransistors.

Field-effect Transistor (FET)

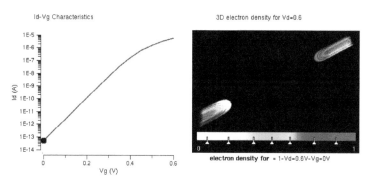

Operation of a FET and its Id-Vg curve. At first, when no gate voltage is applied. There is no inversion electron in the channel, the device is OFF. As gate voltage increase, inversion electron density in the channel increase, current increase, the device turns on.

The *field-effect transistor*, sometimes called a *unipolar transistor*, uses either electrons (in *n-channel FET*) or holes (in *p-channel FET*) for conduction. The four terminals of the FET are named *source*, *gate*, *drain*, and *body* (*substrate*). On most FETs, the body is connected to the source inside the package, and this will be assumed for the following description.

In a FET, the drain-to-source current flows via a conducting channel that connects the *source* region to the *drain* region. The conductivity is varied by the electric field that is produced when a voltage is applied between the gate and source terminals; hence the current flowing between the drain and source is controlled by the voltage applied between the gate and source. As the gate–source voltage (V_{GS}) is increased, the drain–source current (I_{DS}) increases exponentially for V_{GS} below threshold, and then at a roughly quadratic rate ($I_{GS} \square (V_{GS} - V_T)^2$) (where V_T is the threshold voltage at which drain current begins) in the "space-charge-limited" region above threshold. A quadratic behavior is not observed in modern devices, for example, at the 65 nm technology node.

For low noise at narrow bandwidth the higher input resistance of the FET is advantageous.

FETs are divided into two families: *junction FET* (JFET) and *insulated gate FET* (IGFET). The IGFET is more commonly known as a *metal–oxide–semiconductor FET* (MOSFET), reflecting its original construction from layers of metal (the gate), oxide (the insulation), and semiconductor. Unlike IGFETs, the JFET gate forms a p–n diode with the channel which lies between the source and drain. Functionally, this makes the n-channel JFET the solid-state equivalent of the vacuum tube triode which, similarly, forms a diode between its grid and cathode. Also, both devices operate in the *depletion mode*, they both have a high input impedance, and they both conduct current under the control of an input voltage.

Metal–semiconductor FETs (MESFETs) are JFETs in which the reverse biased p–n junction is replaced by a metal–semiconductor junction. These, and the HEMTs (high-electron-mobility transistors, or HFETs), in which a two-dimensional electron gas with very high carrier mobility is used for charge transport, are especially suitable for use at very high frequencies (microwave frequencies; several GHz).

FETs are further divided into *depletion-mode* and *enhancement-mode* types, depending on whether the channel is turned on or off with zero gate-to-source voltage. For enhancement mode, the channel is off at zero bias, and a gate potential can "enhance" the conduction. For the depletion mode, the channel is on at zero bias, and a gate potential (of the opposite polarity) can "deplete" the channel, reducing conduction. For either mode, a more positive gate voltage corresponds to a higher current for n-channel devices and a lower current for p-channel devices. Nearly all JFETs are depletion-mode because the diode junctions would forward bias and conduct if they were enhancement-mode devices; most IGFETs are enhancement-mode types.

Usage of Bipolar and Field-effect Transistors

The bipolar junction transistor (BJT) was the most commonly used transistor in the 1960s and 70s. Even after MOSFETs became widely available, the BJT remained the transistor of choice for many analog circuits such as amplifiers because of their greater linearity and ease of manufacture. In integrated circuits, the desirable properties of MOSFETs allowed them to capture nearly all market share for digital circuits. Discrete MOSFETs can be applied in transistor applications, including analog circuits, voltage regulators, amplifiers, power transmitters and motor drivers.

Other Transistor Types

Transistor symbol created on Portuguese pavement in the University of Aveiro.

- Bipolar junction transistor (BJT):

  - heterojunction bipolar transistor, up to several hundred GHz, common in modern ultrafast and RF circuits;

  - Schottky transistor;

  - avalanche transistor:

  - Darlington transistors are two BJTs connected together to provide a high current gain equal to the product of the current gains of the two transistors;

  - insulated-gate bipolar transistors (IGBTs) use a medium-power IGFET, similarly connected to a power BJT, to give a high input impedance. Power diodes are often connected between certain terminals depending on specific use. IGBTs are particularly suitable for heavy-duty industrial applications. The ASEA Brown Boveri (ABB) *5SNA2400E170100* illustrates just how far power semiconductor technology has advanced. Intended for three-phase power supplies, this device houses three n–p–n IGBTs in a case measuring 38 by 140 by 190 mm and weighing 1.5 kg. Each IGBT is rated at 1,700 volts and can handle 2,400 amperes;

- phototransistor;

- multiple-emitter transistor, used in transistor–transistor logic and integrated current mirrors;

- multiple-base transistor, used to amplify very-low-level signals in noisy environments such as the pickup of a record player or radio front ends. Effectively, it is a very large number of transistors in parallel where, at the output, the signal is added constructively, but random noise is added only stochastically.

- Field-effect transistor (FET):

    - carbon nanotube field-effect transistor (CNFET), where the channel material is replaced by a carbon nanotube;

    - junction gate field-effect transistor (JFET), where the gate is insulated by a reverse-biased p–n junction;

    - metal–semiconductor field-effect transistor (MESFET), similar to JFET with a Schottky junction instead of a p–n junction;

        - high-electron-mobility transistor (HEMT);

    - metal–oxide–semiconductor field-effect transistor (MOSFET), where the gate is insulated by a shallow layer of insulator;

    - inverted-T field-effect transistor (ITFET);

    - fin field-effect transistor (FinFET), source/drain region shapes fins on the silicon surface;

    - fast-reverse epitaxial diode field-effect transistor (FREDFET);

    - thin-film transistor, in LCDs;

    - organic field-effect transistor (OFET), in which the semiconductor is an organic compound;

    - ballistic transistor;

    - floating-gate transistor, for non-volatile storage;

    - FETs used to sense environment;

        - ion-sensitive field-effect transistor (IFSET), to measure ion concentrations in solution,

        - electrolyte–oxide–semiconductor field-effect transistor (EOSFET), neurochip,

        - deoxyribonucleic acid field-effect transistor (DNAFET).

- Tunnel field-effect transistor, where it switches by modulating quantum tunnelling through a barrier.

- Diffusion transistor, formed by diffusing dopants into semiconductor substrate; can be both BJT and FET.

- Unijunction transistor, can be used as simple pulse generators. It comprise a main body of either P-type or N-type semiconductor with ohmic contacts at each end (terminals *Base1* and *Base2*). A junction with the opposite semiconductor type is formed at a point along the length of the body for the third terminal (*Emitter*).

- Single-electron transistors (SET), consist of a gate island between two tunneling junctions. The tunneling current is controlled by a voltage applied to the gate through a capacitor.

- Nanofluidic transistor, controls the movement of ions through sub-microscopic, water-filled channels.

- Multigate devices:

  - tetrode transistor;

  - pentode transistor;

  - trigate transistor (prototype by Intel);

  - dual-gate field-effect transistors have a single channel with two gates in cascode; a configuration optimized for *high-frequency amplifiers*, *mixers*, and oscillators.

- Junctionless nanowire transistor (JNT), uses a simple nanowire of silicon surrounded by an electrically isolated "wedding ring" that acts to gate the flow of electrons through the wire.

- Vacuum-channel transistor, when in 2012, NASA and the National Nanofab Center in South Korea were reported to have built a prototype vacuum-channel transistor in only 150 nanometers in size, can be manufactured cheaply using standard silicon semiconductor processing, can operate at high speeds even in hostile environments, and could consume just as much power as a standard transistor.

- Organic electrochemical transistor.

Part Numbering Standards/Specifications

The types of some transistors can be parsed from the part number. There are three major semiconductor naming standards; in each the alphanumeric prefix provides clues to type of the device.

Japanese Industrial Standard (JIS)

JIS Transistor Prefix Table	
Prefix	Type of transistor
2SA	high-frequency p–n–p BJTs
2SB	audio-frequency p–n–p BJTs

2SC	high-frequency n–p–n BJTs
2SD	audio-frequency n–p–n BJTs
2SJ	P-channel FETs (both JFETs and MOSFETs)
2SK	N-channel FETs (both JFETs and MOSFETs)

The *JIS-C-7012* specification for transistor part numbers starts with "2S", e.g. 2SD965, but sometimes the "2S" prefix is not marked on the package – a 2SD965 might only be marked "D965"; a 2SC1815 might be listed by a supplier as simply "C1815". This series sometimes has suffixes (such as "R", "O", "BL", standing for "red", "orange", "blue", etc.) to denote variants, such as tighter h_{FE} (gain) groupings.

European Electronic Component Manufacturers Association (EECA)

The Pro Electron standard, the European Electronic Component Manufacturers Association part numbering scheme, begins with two letters: the first gives the semiconductor type (A for germanium, B for silicon, and C for materials like GaAs); the second letter denotes the intended use (A for diode, C for general-purpose transistor, etc.). A 3-digit sequence number (or one letter then 2 digits, for industrial types) follows. With early devices this indicated the case type. Suffixes may be used, with a letter (e.g. "C" often means high h_{FE}, such as in: BC549C) or other codes may follow to show gain (e.g. BC327-25) or voltage rating (e.g. BUK854-800A). The more common prefixes are:

Pro Electron / EECA Transistor Prefix Table					
Prefix class	Type and usage	Example	Equivalent	Reference	
AC	Germanium small-signal AF transistor	AC126	NTE102A	Datasheet	
AD	Germanium AF power transistor	AD133	NTE179	Datasheet	
AF	Germanium small-signal RF transistor	AF117	NTE160	Datasheet	
AL	Germanium RF power transistor	ALZ10	NTE100	Datasheet	
AS	Germanium switching transistor	ASY28	NTE101	Datasheet	
AU	Germanium power switching transistor	AU103	NTE127	Datasheet	
BC	Silicon, small-signal transistor ("general purpose")	BC548	2N3904	Datasheet	
BD	Silicon, power transistor	BD139	NTE375	Datasheet	
BF	Silicon, RF (high frequency) BJT or FET	BF245	NTE133	Datasheet	
BS	Silicon, switching transistor (BJT or MOSFET)	BS170	2N7000	Datasheet	
BL	Silicon, high frequency, high power (for transmitters)	BLW60	NTE325	Datasheet	
BU	Silicon, high voltage (for CRT horizontal deflection circuits)	BU2520A	NTE2354	Datasheet	
CF	Gallium Arsenide small-signal Microwave transistor (MESFET)	CF739	–	Datasheet	
CL	Gallium Arsenide Microwave power transistor (FET)	CLY10	–	Datasheet	

Joint Electron Devices Engineering Council (JEDEC)

The JEDEC *EIA370* transistor device numbers usually start with "2N", indicating a three-terminal device (dual-gate field-effect transistors are four-terminal devices, so begin with 3N), then a 2, 3 or 4-digit sequential number with no significance as to device properties (although early devices with low numbers tend to be germanium). For example, 2N3055 is a silicon n–p–n power transistor, 2N1301 is a p–n–p germanium switching transistor. A letter suffix (such as "A") is sometimes used to indicate a newer variant, but rarely gain groupings.

Proprietary

Manufacturers of devices may have their own proprietary numbering system, for example CK722. Since devices are second-sourced, a manufacturer's prefix (like "MPF" in MPF102, which originally would denote a Motorola FET) now is an unreliable indicator of who made the device. Some proprietary naming schemes adopt parts of other naming schemes, for example a PN2222A is a (possibly Fairchild Semiconductor) 2N2222A in a plastic case (but a PN108 is a plastic version of a BC108, not a 2N108, while the PN100 is unrelated to other xx100 devices).

Military part numbers sometimes are assigned their own codes, such as the British Military CV Naming System.

Manufacturers buying large numbers of similar parts may have them supplied with "house numbers", identifying a particular purchasing specification and not necessarily a device with a standardized registered number. For example, an HP part 1854,0053 is a (JEDEC) 2N2218 transistor which is also assigned the CV number: CV7763

Naming Problems

With so many independent naming schemes, and the abbreviation of part numbers when printed on the devices, ambiguity sometimes occurs. For example, two different devices may be marked "J176" (one the J176 low-power JFET, the other the higher-powered MOSFET 2SJ176).

As older "through-hole" transistors are given surface-mount packaged counterparts, they tend to be assigned many different part numbers because manufacturers have their own systems to cope with the variety in pinout arrangements and options for dual or matched n–p–n+p–n–p devices in one pack. So even when the original device (such as a 2N3904) may have been assigned by a standards authority, and well known by engineers over the years, the new versions are far from standardized in their naming.

Construction

Semiconductor Material

Semiconductor material characteristics				
Semiconductor material	Junction forward voltage V @ 25 °C	Electron mobility m²/(V·s) @ 25 °C	Hole mobility m²/(V·s) @ 25 °C	Max. junction temp. °C
Ge	0.27	0.39	0.19	70 to 100

Si	0.71	0.14	0.05	150 to 200
GaAs	1.03	0.85	0.05	150 to 200
Al-Si junction	0.3	—	—	150 to 200

The first BJTs were made from germanium (Ge). Silicon (Si) types currently predominate but certain advanced microwave and high-performance versions now employ the *compound semiconductor* material gallium arsenide (GaAs) and the *semiconductor alloy* silicon germanium (SiGe). Single element semiconductor material (Ge and Si) is described as *elemental.*

Rough parameters for the most common semiconductor materials used to make transistors are given in the table to the right; these parameters will vary with increase in temperature, electric field, impurity level, strain, and sundry other factors.

The *junction forward voltage* is the voltage applied to the emitter–base junction of a BJT in order to make the base conduct a specified current. The current increases exponentially as the junction forward voltage is increased. The values given in the table are typical for a current of 1 mA (the same values apply to semiconductor diodes). The lower the junction forward voltage the better, as this means that less power is required to "drive" the transistor. The junction forward voltage for a given current decreases with increase in temperature. For a typical silicon junction the change is -2.1 mV/°C. In some circuits special compensating elements (sensistors) must be used to compensate for such changes.

The density of mobile carriers in the channel of a MOSFET is a function of the electric field forming the channel and of various other phenomena such as the impurity level in the channel. Some impurities, called dopants, are introduced deliberately in making a MOSFET, to control the MOSFET electrical behavior.

The *electron mobility* and *hole mobility* columns show the average speed that electrons and holes diffuse through the semiconductor material with an electric field of 1 volt per meter applied across the material. In general, the higher the electron mobility the faster the transistor can operate. The table indicates that Ge is a better material than Si in this respect. However, Ge has four major shortcomings compared to silicon and gallium arsenide:

- Its maximum temperature is limited;
- it has relatively high leakage current;
- it cannot withstand high voltages;
- it is less suitable for fabricating integrated circuits.

Because the electron mobility is higher than the hole mobility for all semiconductor materials, a given bipolar n–p–n transistor tends to be swifter than an equivalent p–n–p transistor. GaAs has the highest electron mobility of the three semiconductors. It is for this reason that GaAs is used in high-frequency applications. A relatively recent FET development, the *high-electron-mobility transistor* (HEMT), has a heterostructure (junction between different semiconductor materials) of aluminium gallium arsenide (AlGaAs)-gallium arsenide (GaAs) which has twice the electron mobility of a GaAs-metal barrier junction. Because of their high speed and low noise, HEMTs are used in satellite receivers working at frequencies around 12 GHz. HEMTs based on gallium nitride and aluminium gallium nitride (AlGaN/GaN HEMTs) provide a still higher electron mobility and are being developed for various applications.

Max. junction temperature values represent a cross section taken from various manufacturers' data sheets. This temperature should not be exceeded or the transistor may be damaged.

Al–Si junction refers to the high-speed (aluminum–silicon) metal–semiconductor barrier diode, commonly known as a Schottky diode. This is included in the table because some silicon power IGFETs have a *parasitic* reverse Schottky diode formed between the source and drain as part of the fabrication process. This diode can be a nuisance, but sometimes it is used in the circuit.

Packaging

Assorted discrete transistors.

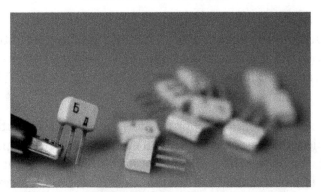

Soviet KT315b transistors.

Discrete transistors are individually packaged transistors. Transistors come in many different semiconductor packages. The two main categories are *through-hole* (or *leaded*), and *surface-mount*, also known as *surface-mount device* (SMD). The *ball grid array* (BGA) is the latest surface-mount package (currently only for large integrated circuits). It has solder "balls" on the underside in place of leads. Because they are smaller and have shorter interconnections, SMDs have better high-frequency characteristics but lower power rating.

Transistor packages are made of glass, metal, ceramic, or plastic. The package often dictates the power rating and frequency characteristics. Power transistors have larger packages that can be clamped to heat sinks for enhanced cooling. Additionally, most power transistors have the collector or drain physically connected to the metal enclosure. At the other extreme, some surface-mount *microwave* transistors are as small as grains of sand.

Often a given transistor type is available in several packages. Transistor packages are mainly standardized, but the assignment of a transistor's functions to the terminals is not: other transistor types can assign other functions to the package's terminals. Even for the same transistor type the terminal assignment can vary (normally indicated by a suffix letter to the part number, q.e. BC212L and BC212K).

Nowadays most transistors come in a wide range of SMT packages, in comparison the list of available through-hole packages is relatively small, here is a short list of the most common through-hole transistors packages in alphabetical order: ATV, E-line, MRT, HRT, SC-43, SC-72, TO-3, TO-18, TO-39, TO-92, TO-126, TO220, TO247, TO251, TO262, ZTX851

Flexible Transistors

Researchers have made several kinds of flexible transistors, including organic field-effect transistors. Flexible transistors are useful in some kinds of flexible displays and other flexible electronics.

Voltmeter

Demonstration voltmeter

A voltmeter is an instrument used for measuring electrical potential difference between two points in an electric circuit. Analog voltmeters move a pointer across a scale in proportion to the voltage of the circuit; digital voltmeters give a numerical display of voltage by use of an analog to digital converter.

A voltmeter in a circuit diagram is represented by the letter V in a circle.

Voltmeters are made in a wide range of styles. Instruments permanently mounted in a panel are used to monitor generators or other fixed apparatus. Portable instruments, usually equipped to also measure current and resistance in the form of a multimeter, are standard test instruments used in electrical and electronics work. Any measurement that can be converted to a voltage can be displayed on a meter that is suitably calibrated; for example, pressure, temperature, flow or level in a chemical process plant.

General purpose analog voltmeters may have an accuracy of a few percent of full scale, and are used with voltages from a fraction of a volt to several thousand volts. Digital meters can be made with high accuracy, typically better than 1%. Specially calibrated test instruments have higher accuracies, with laboratory instruments capable of measuring to accuracies of a few parts per million. Meters using amplifiers can measure tiny voltages of microvolts or less.

Part of the problem of making an accurate voltmeter is that of calibration to check its accuracy. In laboratories, the Weston Cell is used as a standard voltage for precision work. Precision voltage references are available based on electronic circuits.

Analog Voltmeter

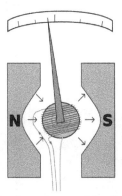

A moving coil galvanometer of the d'Arsonval type.

- The red wire carries the current to be measured.
- The restoring spring is shown in green.
- N and S are the north and south poles of the magnet.

A moving coil galvanometer can be used as a voltmeter by inserting a resistor in series with the instrument. The galvanometer has a coil of fine wire suspended in a strong magnetic field. When an electric current is applied, the interaction of the magnetic field of the coil and of the stationary magnet creates a torque, tending to make the coil rotate. The torque is proportional to the current through the coil. The coil rotates, compressing a spring that opposes the rotation. The deflection of the coil is thus proportional to the current, which in turn is proportional to the applied voltage, which is indicated by a pointer on a scale.

One of the design objectives of the instrument is to disturb the circuit as little as possible and so the instrument should draw a minimum of current to operate. This is achieved by using a sensitive galvanometer in series with a high resistance.

The sensitivity of such a meter can be expressed as "ohms per volt", the number of ohms resistance in the meter circuit divided by the full scale measured value. For example, a meter with a sensitivity of 1000 ohms per volt would draw 1 milliampere at full scale voltage; if the full scale was 200 volts, the resistance at the instrument's terminals would be 200,000 ohms and at full scale the meter would draw 1 milliampere from the circuit under test. For multi-range instruments, the input resistance varies as the instrument is switched to different ranges.

Moving-coil instruments with a permanent-magnet field respond only to direct current. Measurement of AC voltage requires a rectifier in the circuit so that the coil deflects in only one direction. Some moving-coil instruments are also made with the zero position in the middle of the scale instead of at one end; these are useful if the voltage reverses its polarity.

Voltmeters operating on the electrostatic principle use the mutual repulsion between two charged plates to deflect a pointer attached to a spring. Meters of this type draw negligible current but are sensitive to voltages over about 100 volts and work with either alternating or direct current.

VTVMs and FET-VMs

The sensitivity and input resistance of a voltmeter can be increased if the current required to deflect the meter pointer is supplied by an amplifier and power supply instead of by the circuit under test. The electronic amplifier between input and meter gives two benefits; a rugged moving coil instrument can be used, since its sensitivity need not be high, and the input resistance can be made high, reducing the current drawn from the circuit under test. Amplified voltmeters often have an input resistance of 1, 10, or 20 megohms which is independent of the range selected. A once-popular form of this instrument used a vacuum tube in the amplifier circuit and so was called the vacuum tube voltmeter, or VTVM. These were almost always powered by the local AC line current and so were not particularly portable. Today these circuits use a solid-state amplifier using field-effect transistors, hence FET-VM, and appear in handheld digital multimeters as well as in bench and laboratory instruments. These are now so ubiquitous that they have largely replaced non-amplified multimeters except in the least expensive price ranges.

Most VTVMs and FET-VMs handle DC voltage, AC voltage, and resistance measurements; modern FET-VMs add current measurements and often other functions as well. A specialized form of the VTVM or FET-VM is the AC voltmeter. These instruments are optimized for measuring AC voltage. They have much wider bandwidth and better sensitivity than a typical multifunction device.

Digital Voltmeter

Two digital voltmeters. Note the 40 microvolt difference between the two measurements, an offset of 34 parts per million.

A digital voltmeter (DVM) measures an unknown input voltage by converting the voltage to a digital value and then displays the voltage in numeric form. DVMs are usually designed around a special type of analog-to-digital converter called an integrating converter.

DVM measurement accuracy is affected by many factors, including temperature, input impedance, and DVM power supply voltage variations. Less expensive DVMs often have input resistance on the order of 10 MΩ. Precision DVMs can have input resistances of 1 GΩ or higher for the lower voltage ranges (e.g. less than 20 V). To ensure that a DVM's accuracy is within the manufacturer's specified tolerances, it must be periodically calibrated against a voltage standard such as the Weston cell.

The first digital voltmeter was invented and produced by Andrew Kay of Non-Linear Systems (and later founder of Kaypro) in 1954.

Electrical Telegraph

A printing electrical telegraph receiver, with transmitter key at bottom right

An electrical telegraph is a telegraph that uses electrical signals, usually conveyed via dedicated telecommunication lines or radio.

The electrical telegraph, or more commonly just telegraph, superseded optical semaphore telegraph systems, such as Claude Chappe's towers designed for communication among the French military, and Friedrich Clemens Gerke for the Prussian military, thus becoming the first form of electrical telecommunications. In a matter of decades after their creation, electrical telegraph networks permitted people and commerce to transmit messages across both continents and oceans almost instantly, with widespread social and economic impacts.

History

Early Work

From early studies of electricity, electrical phenomena were known to travel with great speed, and many experimenters worked on the application of electricity to communications at a distance.

Sömmering's electric telegraph in 1809

All the known effects of electricity - such as sparks, electrostatic attraction, chemical changes, electric shocks, and later electromagnetism - were applied to the problems of detecting controlled transmissions of electricity at various distances.

In 1753 an anonymous writer in the *Scots Magazine* suggested an electrostatic telegraph. Using one wire for each letter of the alphabet, a message could be transmitted by connecting the wire terminals in turn to an electrostatic machine, and observing the deflection of pith balls at the far end. Telegraphs employing electrostatic attraction were the basis of early experiments in electrical telegraphy in Europe, but were abandoned as being impractical and were never developed into a useful communication system.

In 1800 Alessandro Volta invented the voltaic pile, allowing for a continuous current of electricity for experimentation. This became a source of a low-voltage current that could be used to produce more distinct effects, and which was far less limited than the momentary discharge of an electrostatic machine, which with Leyden jars were the only previously known man-made sources of electricity.

Another very early experiment in electrical telegraphy was an 'electrochemical telegraph' created by the German physician, anatomist and inventor Samuel Thomas von Sömmering in 1809, based on an earlier, less robust design of 1804 by Catalan polymath and scientist Francisco Salva Campillo. Both their designs employed multiple wires (up to 35) to represent almost all Latin letters and numerals. Thus, messages could be conveyed electrically up to a few kilometers (in von Sömmering's design), with each of the telegraph receiver's wires immersed in a separate glass tube of acid. An electric current was sequentially applied by the sender through the various wires representing each digit of a message; at the recipient's end the currents electrolysed the acid in the tubes in sequence, releasing streams of hydrogen bubbles next to each associated letter or numeral. The telegraph receiver's operator would watch the bubbles and could then record the transmitted message. This is in contrast to later telegraphs that used a single wire (with ground return).

Hans Christian Ørsted discovered in 1820 that an electric current produces a magnetic field which will deflect a compass needle. In the same year Johann Schweigger invented the galvanometer, with a coil of wire around a compass, which could be used as a sensitive indicator for an electric current. In 1821, André-Marie Ampère suggested that telegraphy could be done by a system of galvanometers, with one wire per galvanometer to indicate each letter, and said he had experimented successfully with such a system. In 1824, Peter Barlow said that such a system only worked to a distance of about 200 feet (61 m), and so was impractical.

In 1825 William Sturgeon invented the electromagnet, with a single winding of uninsulated wire on a piece of varnished iron, which increased the magnetic force produced by electric current. Joseph Henry improved it in 1828 by placing several windings of insulated wire around the bar, creating a much more powerful electromagnet which could operate a telegraph through the high resistance of long telegraph wires. During his tenure at The Albany Academy from 1826 to 1832, Henry first demonstrated the theory of the 'magnetic telegraph' by ringing a bell through a mile of wire strung around the room.

In 1835 Joseph Henry and Edward Davy invented the critical electrical relay. Davy's relay used a magnetic needle which dipped into a mercury contact when an electric current passed through the surrounding coil. This allowed a weak current to switch a larger current to operate a powerful local electromagnet over very long distances.

First Working Systems

The first working telegraph was built by the English inventor Francis Ronalds in 1816 and used static electricity. At the family home on Hammersmith Mall, he set up a complete subterranean system in a 175 yard long trench as well as an eight mile long overhead telegraph. The lines were connected at both ends to clocks marked with the letters of the alphabet and electrical impulses sent along the wire were used to transmit messages. Offering his invention to the Admiralty in July 1816, it was rejected as "wholly unnecessary". His account of the scheme and the possibilities of rapid global communication in *Descriptions of an Electrical Telegraph and of some other Electrical Apparatus* was the first published work on electric telegraphy and even described the risk of signal retardation due to induction. Elements of Ronalds' design were utilised in the subsequent commercialisation of the telegraph over 20 years later.

Pavel Shilling, an early pioneer of electrical telegraphy

The telegraph invented by Baron Schilling von Canstatt in 1832 had a transmitting device which consisted of a keyboard with 16 black-and-white keys. These served for switching the electric current. The receiving instrument consisted of six galvanometers with magnetic needles, suspended from silk threads. Both stations of Shilling's telegraph were connected by eight wires; six were connected with the galvanometers, one served for the return current and one - for a signal bell. When at the starting station the operator pressed a key, the corresponding pointer was deflected at the receiving station. Different positions of black and white flags on different disks gave combinations which corresponded to the letters or numbers. Pavel Shilling subsequently improved its apparatus. He reduced the number of connecting wires from eight to two.

On 21 October 1832, Schilling managed a short-distance transmission of signals between two telegraphs in different rooms of his apartment. In 1836 the British government attempted to buy the design but Schilling instead accepted overtures from Nicholas I of Russia. Schilling's telegraph was tested on a 5-kilometre-long (3.1 mi) experimental underground and underwater cable, laid around the building of the main Admiralty in Saint Petersburg and was approved for a telegraph between the imperial palace at Peterhof and the naval base at Kronstadt. However, the project was cancelled following Schilling's death in 1837. Schilling was also one of the first to put into practice the idea of the binary system of signal transmission.

In 1833, Carl Friedrich Gauss, together with the physics professor Wilhelm Weber in Göttingen installed a 1,200-metre-long (3,900 ft) wire above the town's roofs. Gauss combined the Poggendorff-Schweigger multiplicator with his magnetometer to build a more sensitive device, the galvanometer. To change the direction of the electric current, he constructed a commutator of his own. As a result, he was able to make the distant needle move in the direction set by the commutator on the other end of the line.

At first, they used the telegraph to coordinate time, but soon they developed other signals; finally, their own alphabet. The alphabet was encoded in a binary code which was transmitted by positive or negative voltage pulses which were generated by means of moving an induction coil up and down over a permanent magnet and connecting the coil with the transmission wires by means of the commutator. The page of Gauss' laboratory notebook containing both his code and the first message transmitted, as well as a replica of the telegraph made in the 1850s under the instructions of Weber are kept in the faculty of physics of Göttingen University.

Gauss was convinced that this communication would be a help to his kingdom's towns. Later in the same year, instead of a Voltaic pile, Gauss used an induction pulse, enabling him to transmit seven letters a minute instead of two. The inventors and university were too poor to develop the telegraph on their own, but they received funding from Alexander von Humboldt. Carl August Steinheil in Munich was able to build a telegraph network within the city in 1835-6. He installed a telegraph line along the first German railroad in 1835.

Across the Atlantic, in 1836 an American scientist, Dr. David Alter, invented the first known American electric telegraph, in Elderton, Pennsylvania, one year before the Morse telegraph. Alter demonstrated it to witnesses but never developed the idea into a practical system. He was interviewed later for the book Biographical and Historical Cyclopedia of Indiana and Armstrong Counties, in which he said: "I may say that there is no connection at all between the telegraph of Morse and others and that of myself.... Professor Morse most probably never heard of me or my Elderton telegraph."

Commercial Telegraphy

Cooke and Wheatstone System

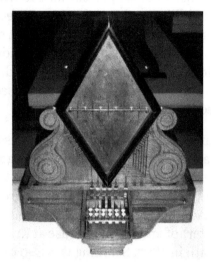

Cooke and Wheatstone's five-needle, six-wire telegraph

The first commercial electrical telegraph, the Cooke and Wheatstone telegraph, was co-developed by William Fothergill Cooke and Charles Wheatstone. In May 1837 they patented a telegraph system which used a number of needles on a board that could be moved to point to letters of the alphabet. The patent recommended a five-needle system, but any number of needles could be used depending on the number of characters it was required to code. A four-needle system was installed between Euston and Camden Town in London on a rail line being constructed by Robert Stephenson between London and Birmingham. It was successfully demonstrated on 25 July 1837. Euston needed to signal to an engine house at Camden Town to start hauling the locomotive up the incline. As at Liverpool, the electric telegraph was in the end rejected in favour of a pneumatic system with whistles.

Cooke and Wheatstone had their first commercial success with a system installed on the Great Western Railway over the 13 miles (21 km) from Paddington station to West Drayton in 1838, the first commercial telegraph in the world. This was a five-needle, six-wire system. The cables were originally installed underground in a steel conduit. However, the cables soon began to fail as a result of deteriorating insulation and were replaced with uninsulated wires on poles. As an interim measure, a two-needle system was used with three of the remaining working underground wires, which despite using only two needles had a greater number of codes. But when the line was extended to Slough in 1843, a one-needle, two-wire system was installed.

From this point the use of the electric telegraph started to grow on the new railways being built from London. The London and Blackwall Railway (another rope-hauled application) was equipped with the Cooke and Wheatstone telegraph when it opened in 1840, and many others followed. The one-needle telegraph proved highly successful on British railways, and 15,000 sets were still in use at the end of the nineteenth century. Some remained in service in the 1930s. In September 1845 the financier John Lewis Ricardo and Cooke formed the Electric Telegraph Company, the first public telegraphy company in the world. This company bought out the Cooke and Wheatstone patents and solidly established the telegraph business.

As well as the rapid expansion of the use of the telegraphs along the railways, they soon spread into the field of mass communication with the instruments being installed in post offices across the country. The era of mass personal communication had begun.

Morse System

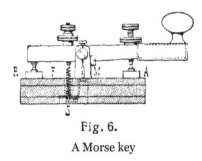

Fig. 6.

A Morse key

An electrical telegraph was independently developed and patented in the United States in 1837 by Samuel Morse. His assistant, Alfred Vail, developed the Morse code signalling alphabet with Morse. The first telegram in the United States was sent by Morse on 11 January 1838, across two miles (3 km) of wire at Speedwell Ironworks near Morristown, New Jersey, although it was only later, in 1844, that he sent the message "WHAT HATH GOD WROUGHT" from the Capitol in Washington to the old Mt. Clare Depot in Baltimore. The Morse/Vail telegraph was quickly deployed in the following two decades; the overland telegraph connected the west coast of the continent to the east coast by 24 October 1861, bringing an end to the Pony Express.

Edward Davy demonstrated his telegraph system in Regent's Park in 1837 and was granted a patent on 4 July 1838. He also developed an electric relay.

Telegraphic Improvements

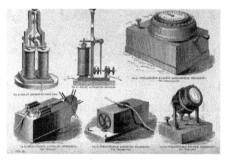

Wheatstone automated telegraph network equipment

A continuing goal in telegraphy was to reduce the cost per message by reducing hand-work, or increasing the sending rate. There were many experiments with moving pointers, and various electrical encodings. However, most systems were too complicated and unreliable. A successful expedient to increase the sending rate was the development of telegraphese.

The first system that didn't require skilled technicians to operate, was Charles Wheatstone's ABC system in 1840 where the letters of the alphabet were arranged around a clock-face, and the signal caused a needle to indicate the letter. This early system required the receiver to be present in real time to record the message and it reached speeds of up to 15 words a minute.

In 1846, Alexander Bain patented a chemical telegraph in Edinburgh. The signal current moved an iron pen across a moving paper tape soaked in a mixture of ammonium nitrate and potassium ferrocyanide, decomposing the chemical and producing readable blue marks in Morse code. The speed of the printing telegraph was 1000 words per minute, but messages still required translation into English by live copyists. Chemical telegraphy came to an end in the US in 1851, when the Morse group defeated the Bain patent in the US District Court.

For a brief period, starting with the New York-Boston line in 1848, some telegraph networks began to employ sound operators, who were trained to understand Morse code aurally. Gradually, the use of sound operators eliminated the need for telegraph receivers to include register and tape. Instead, the receiving instrument was developed into a "sounder," an electromagnet that was energized by a current and attracted a small iron lever. When the sounding key was opened or closed, the sounder lever struck an anvil. The Morse operator distinguished a dot and a dash by the short or long interval between the two clicks. The message was then written out in long-hand.

Royal Earl House developed and patented a letter-printing telegraph system in 1846 which employed an alphabetic keyboard for the transmitter and automatically printed the letters on paper at the receiver, and followed this up with a steam-powered version in 1852. Advocates of printing telegraphy said it would eliminate Morse operators' errors. The House machine was used on four main American telegraph lines by 1852. The speed of the House machine was announced as 2600 words an hour.

A Baudot keyboard, 1884

David Edward Hughes invented the printing telegraph in 1855; it used a keyboard of 26 keys for the alphabet and a spinning type wheel that determined the letter being transmitted by the length of time that had elapsed since the previous transmission. The system allowed for automatic recording on the receiving end. The system was very stable and accurate and became the accepted around the world.

The next improvement was the Baudot code of 1874. French engineer Émile Baudot patented a printing telegraph in which the signals were translated automatically into typographic characters. Each character was assigned a unique code based on the sequence of just five contacts. Operators had to maintain a steady rhythm, and the usual speed of operation was 30 words per minute.

By this point reception had been automated, but the speed and accuracy of the transmission was still limited to the skill of the human operator. The first practical automated system was patented by Charles Wheatstone, the original inventor of the telegraph. The message (in Morse code) was typed onto a piece of perforated tape using a keyboard-like device called the 'Stick Punch'. The transmitter automatically ran the tape through and transmitted the message at the then exceptionally high speed of 70 words per minute.

Teleprinters

Phelps' Electro-motor Printing Telegraph from circa 1880, the last and most advanced telegraphy mechanism designed by George May Phelps

An early successful teleprinter was invented by Frederick G. Creed. In Glasgow he created his first keyboard perforator, which used compressed air to punch the holes. He also created a reperforator (receiving perforator) and a printer. The reperforator punched incoming Morse signals on to paper tape and the printer decoded this tape to produce alphanumeric characters on plain paper. This was the origin of the Creed High Speed Automatic Printing System, which could run at an unprecedented 200 words per minute. His system was adopted by the *Daily Mail* for daily transmission of the newspaper contents.

By the 1930s teleprinters were being produced by Teletype in the US, Creed in Britain and Siemens in Germany.

With the invention of the teletypewriter, telegraphic encoding became fully automated. Early teletypewriters used the ITA-1 Baudot code, a five-bit code. This yielded only thirty-two codes, so it was over-defined into two "shifts", "letters" and "figures". An explicit, unshared shift code prefaced each set of letters and figures.

A Siemens T100 Telex machine

By 1935, message routing was the last great barrier to full automation. Large telegraphy providers began to develop systems that used telephone-like rotary dialling to connect teletypewriters. These machines were called "Telex" (TELegraph EXchange). Telex machines first performed rotary-telephone-style pulse dialling for circuit switching, and then sent data by ITA2. This "type A" Telex routing functionally automated message routing.

The first wide-coverage Telex network was implemented in Germany during the 1930s as a network used to communicate within the government.

At the rate of 45.45 (±0.5%) baud — considered speedy at the time — up to 25 telex channels could share a single long-distance telephone channel by using *voice frequency telegraphy multiplexing*, making telex the least expensive method of reliable long-distance communication.

Automatic teleprinter exchange service was introduced into Canada by CPR Telegraphs and CN Telegraph in July 1957 and in 1958, Western Union started to build a Telex network in the United States.

Oceanic Telegraph Cables

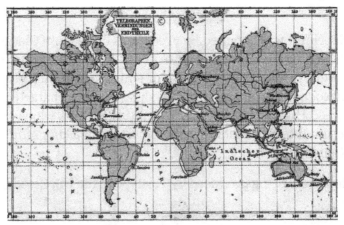

Major telegraph lines in 1891.

Soon after the first successful telegraph systems were operational, the possibility of transmitting messages across the sea by way of submarine communications cables was first mooted. One of the primary technical challenges was to sufficiently insulate the submarine cable to prevent the current from leaking out into the water. In 1842, a Scottish surgeon William Montgomerie introduced gutta-percha, the adhesive juice of the *Palaquium gutta* tree, to Europe. Michael Faraday and Wheatstone soon discovered the merits of gutta-percha as an insulator, and in 1845, the latter suggested that it should be employed to cover the wire which was proposed to be laid from Dover to Calais. It was tried on a wire laid across the Rhine between Deutz and Cologne. In 1849, C.V. Walker, electrician to the South Eastern Railway, submerged a two-mile wire coated with gutta-percha off the coast from Folkestone, which was tested successfully.

John Watkins Brett, an engineer from Bristol, sought and obtained permission from Louis-Philippe in 1847 to establish telegraphic communication between France and England. The first undersea cable was laid in 1850, connecting the two countries and was followed by connections to Ireland and the Low Countries.

The Atlantic Telegraph Company was formed in London in 1856 to undertake to construct a commercial telegraph cable across the Atlantic ocean. It was successfully completed on 18 July 1866 by the ship SS *Great Eastern*, captained by Sir James Anderson after many mishaps along the away. Earlier transatlantic submarine cables installations were attempted in 1857, 1858 and 1865. The 1857 cable only operated intermittently for a few days or weeks before it failed. The study of underwater telegraph cables accelerated interest in mathematical analysis of very long transmission lines. The telegraph lines from Britain to India were connected in 1870 (those several companies combined to form the *Eastern Telegraph Company* in 1872).

Australia was first linked to the rest of the world in October 1872 by a submarine telegraph cable at Darwin. This brought news reportage from the rest of the world. The telegraph across the Pacific was completed in 1902, finally encircling the world.

From the 1850s until well into the 20th century, British submarine cable systems dominated the world system. This was set out as a formal strategic goal, which became known as the All Red Line. In 1896, there were thirty cable laying ships in the world and twenty-four of them were owned by British companies. In 1892, British companies owned and operated two-thirds of the world's cables and by 1923, their share was still 42.7 percent. During World War I, Britain's telegraph communications were almost completely uninterrupted, while it was able to quickly cut Germany's cables worldwide.

End of the Telegraph Era

In the United States, Western Union discontinued all telegram and commercial messaging services on 27 January 2006, although it still offered its electronic money transfer services.

India's state-owned telecom company, BSNL, ended its telegraph service on 14 July 2013. It was reportedly the world's last existing true electric telegraph system.

Electric Motor

Various electric motors, compared to 9 V battery.

An electric motor is an electrical machine that converts electrical energy into mechanical energy. The reverse of this would be the conversion of mechanical energy into electrical energy and is done by an electric generator.

In normal motoring mode, most electric motors operate through the interaction between an electric motor's magnetic field and winding currents to generate force within the motor. In certain applications, such as in the transportation industry with traction motors, electric motors can operate in both motoring and generating or braking modes to also produce electrical energy from mechanical energy.

Found in applications as diverse as industrial fans, blowers and pumps, machine tools, household appliances, power tools, and disk drives, electric motors can be powered by direct current (DC) sources, such as from batteries, motor vehicles or rectifiers, or by alternating current (AC) sources, such as from the power grid, inverters or generators. Small motors may be found in electric watches. General-purpose motors with highly standardized dimensions and characteristics provide convenient mechanical power for industrial use. The largest of electric motors are used for ship propulsion, pipeline compression and pumped-storage applications with ratings reaching 100 megawatts. Electric motors may be classified by electric power source type, internal construction, application, type of motion output, and so on.

Electric motors are used to produce linear or rotary force (torque), and should be distinguished from devices such as magnetic solenoids and loudspeakers that convert electricity into motion but do not generate usable mechanical powers, which are respectively referred to as actuators and transducers.

Cutaway view through stator of induction motor.

History

Early Motors

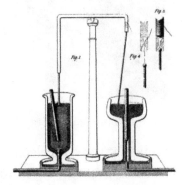

Faraday's electromagnetic experiment, 1821

Perhaps the first electric motors were simple electrostatic devices created by the Scottish monk Andrew Gordon in the 1740s. The theoretical principle behind production of mechanical force by the interactions of an electric current and a magnetic field, Ampère's force law, was discovered later by André-Marie Ampère in 1820. The conversion of electrical energy into mechanical energy by electromagnetic means was demonstrated by the British scientist Michael Faraday in 1821. A free-hanging wire was dipped into a pool of mercury, on which a permanent magnet (PM) was placed. When a current was passed through the wire, the wire rotated around the magnet, showing that the current gave rise to a close circular magnetic field around the wire. This motor is often demonstrated in physics experiments, brine substituting for toxic mercury. Though Barlow's wheel was an early refinement to this Faraday demonstration, these and similar homopolar motors were to remain unsuited to practical application until late in the century.

Jedlik's "electromagnetic self-rotor", 1827 (Museum of Applied Arts, Budapest). The historic motor still works perfectly today.

In 1827, Hungarian physicist Ányos Jedlik started experimenting with electromagnetic coils. After Jedlik solved the technical problems of the continuous rotation with the invention of the commutator, he called his early devices "electromagnetic self-rotors". Although they were used only for instructional purposes, in 1828 Jedlik demonstrated the first device to contain the three main components of practical DC motors: the stator, rotor and commutator. The device employed no permanent magnets, as the magnetic fields of both the stationary and revolving components were produced solely by the currents flowing through their windings.

Success with DC Motors

After many other more or less successful attempts with relatively weak rotating and reciprocating apparatus the Prussian Moritz von Jacobi created the first real rotating electric motor in May 1834 that actually developed a remarkable mechanical output power. His motor set a world record which was improved only four years later in September 1838 by Jacobi himself. His second motor was powerful enough to drive a boat with 14 people across a wide river. It was not until 1839/40 that other developers worldwide managed to build motors of similar and later also of higher performance.

The first commutator DC electric motor capable of turning machinery was invented by the British scientist William Sturgeon in 1832. Following Sturgeon's work, a commutator-type direct-current electric motor made with the intention of commercial use was built by the American inventor Thomas Davenport, which he patented in 1837. The motors ran at up to 600 revolutions per minute, and powered machine tools and a printing press. Due to the high cost of primary battery power, the motors were commercially unsuccessful and Davenport went bankrupt. Several inventors followed Sturgeon in the development of DC motors but all encountered the same battery power cost issues. No electricity distribution had been developed at the time. Like Sturgeon's motor, there was no practical commercial market for these motors.

In 1855, Jedlik built a device using similar principles to those used in his electromagnetic self-rotors that was capable of useful work. He built a model electric vehicle that same year.

A major turning point in the development of DC machines took place in 1864, when Antonio Pacinotti described for the first time the ring armature with its symmetrically grouped coils closed upon themselves and connected to the bars of a commutator, the brushes of which delivered practically non-fluctuating current. The first commercially successful DC motors followed the invention by Zénobe Gramme who, in 1871, reinvented Pacinotti's design. In 1873, Gramme showed that his dynamo could be used as a motor, which he demonstrated to great effect at exhibitions in Vienna and Philadelphia by connecting two such DC motors at a distance of up to 2 km away from each other, one as a generator.

In 1886, Frank Julian Sprague invented the first practical DC motor, a non-sparking motor that maintained relatively constant speed under variable loads. Other Sprague electric inventions about this time greatly improved grid electric distribution (prior work done while employed by Thomas Edison), allowed power from electric motors to be returned to the electric grid, provided for electric distribution to trolleys via overhead wires and the trolley pole, and provided controls systems for electric operations. This allowed Sprague to use electric motors to invent the first electric trolley system in 1887–88 in Richmond VA, the electric elevator and control system in 1892, and the electric subway with independently powered centrally controlled cars, which were first installed in 1892 in Chicago by the South Side Elevated Railway where it became popularly known as the "L". Sprague's motor and related inventions led to an explosion of interest and use in electric motors for industry, while almost simultaneously another great inventor was developing its primary competitor, which would become much more widespread. The development of electric motors of acceptable efficiency was delayed for several decades by failure to recognize the extreme importance of a relatively small air gap between rotor and stator. Efficient designs have a comparatively small air gap. [a] The St. Louis motor, long used in classrooms to illustrate motor principles, is extremely inefficient for the same reason, as well as appearing nothing like a modern motor.

Application of electric motors revolutionized industry. Industrial processes were no longer limited by power transmission using line shafts, belts, compressed air or hydraulic pressure. Instead every machine could be equipped with its own electric motor, providing easy control at the point of use, and improving power transmission efficiency. Electric motors applied in agriculture eliminated human and animal muscle power from such tasks as handling grain or pumping water. Household uses of electric motors reduced heavy labor in the home and made higher standards of convenience, comfort and safety possible. Today, electric motors stand for more than half of the electric energy consumption in the US.

Emergence of AC Motors

In 1824, the French physicist François Arago formulated the existence of rotating magnetic fields, termed Arago's rotations, which, by manually turning switches on and off, Walter Baily demonstrated in 1879 as in effect the first primitive induction motor. In the 1880s, many inventors were trying to develop workable AC motors because AC's advantages in long-distance high-voltage transmission were counterbalanced by the inability to operate motors on AC. The first alternating-current commutatorless induction motors were independently invented by Galileo Ferraris and Nikola Tesla, a working motor model having been demonstrated by the former in 1885 and by the latter in 1887. In 1888, the *Royal Academy of Science of Turin* published Ferraris's research detailing the foundations of motor operation while however concluding that "the apparatus based on that principle could not be of any commercial importance as motor."

In 1888, Tesla presented his paper *A New System for Alternating Current Motors and Transformers* to the AIEE that described three patented two-phase four-stator-pole motor types: one with a four-pole rotor forming a non-self-starting reluctance motor, another with a wound rotor forming a self-starting induction motor, and the third a true synchronous motor with separately excited DC supply to rotor winding.

One of the patents Tesla filed in 1887, however, also described a shorted-winding-rotor induction motor. George Westinghouse promptly bought Tesla's patents, employed Tesla to develop them, and assigned C. F. Scott to help Tesla; however, Tesla left for other pursuits in 1889. The constant speed AC induction motor was found not to be suitable for street cars, but Westinghouse engineers successfully adapted it to power a mining operation in Telluride, Colorado in 1891.

Steadfast in his promotion of three-phase development, Mikhail Dolivo-Dobrovolsky invented the three-phase cage-rotor induction motor in 1889 and the three-limb transformer in 1890. This type of motor is now used for the vast majority of commercial applications. However, he claimed that Tesla's motor was not practical because of two-phase pulsations, which prompted him to persist in his three-phase work. Although Westinghouse achieved its first practical induction motor in 1892 and developed a line of polyphase 60 hertz induction motors in 1893, these early Westinghouse motors were two-phase motors with wound rotors until B. G. Lamme developed a rotating bar winding rotor.

The General Electric Company began developing three-phase induction motors in 1891. By 1896, General Electric and Westinghouse signed a cross-licensing agreement for the bar-winding-rotor design, later called the squirrel-cage rotor. Induction motor improvements flowing from these inventions and innovations were such that a 100 horsepower (HP) induction motor currently has the same mounting dimensions as a 7.5 HP motor in 1897.

Motor Construction

Electric motor rotor (left) and stator (right)

Rotor

In an electric motor the moving part is the rotor which turns the shaft to deliver the mechanical power. The rotor usually has conductors laid into it which carry currents that interact with the magnetic field of the stator to generate the forces that turn the shaft. However, some rotors carry permanent magnets, and the stator holds the conductors.

Stator

The stator is the stationary part of the motor's electromagnetic circuit and usually consists of either windings or permanent magnets. The stator core is made up of many thin metal sheets, called laminations. Laminations are used to reduce energy losses that would result if a solid core were used.

Air Gap

The distance between the rotor and stator is called the air gap. The air gap has important effects, and is generally as small as possible, as a large gap has a strong negative effect on the performance of an electric motor. It is the main source of the low power factor at which motors operate.The air gap increases the magnetizing current needed. For this reason air gap should be minimum . Very small gaps may pose mechanical problems in addition to noise and losses.

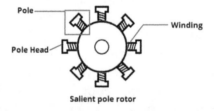

Salient-pole rotor

Windings

Windings are wires that are laid in coils, usually wrapped around a laminated soft iron magnetic core so as to form magnetic poles when energized with current.

Electric machines come in two basic magnet field pole configurations: *salient-pole* machine and *nonsalient-pole* machine. In the salient-pole machine the pole's magnetic field is produced by a winding wound around the pole below the pole face. In the *nonsalient-pole*, or distributed field, or round-rotor, machine, the winding is distributed in pole face slots. A shaded-pole motor has a winding around part of the pole that delays the phase of the magnetic field for that pole.

Some motors have conductors which consist of thicker metal, such as bars or sheets of metal, usually copper, although sometimes aluminum is used. These are usually powered by electromagnetic induction.

Commutator

A commutator is a mechanism used to switch the input of most DC machines and certain AC machines consisting of slip ring segments insulated from each other and from the electric motor's shaft. The motor's armature current is supplied through the stationary brushes in contact with the revolving commutator, which causes required current reversal and applies power to the machine in an optimal manner as the rotor rotates from pole to pole. In absence of such current reversal, the motor would brake to a stop. In light of significant advances in the past few decades due to improved technologies in electronic controller, sensorless control, induction motor, and permanent magnet motor fields, electromechanically commutated motors are increasingly being displaced by externally commutated induction and permanent-magnet motors.

A toy's small DC motor with its commutator

Motor Supply and Control

Motor Supply

A DC motor is usually supplied through slip ring commutator as described above. AC motors' commutation can be either slip ring commutator or externally commutated type, can be fixed-speed or variable-speed control type, and can be synchronous or asynchronous type. Universal motors can run on either AC or DC.

Motor Control

Fixed-speed controlled AC motors are provided with direct-on-line or soft-start starters.

Variable speed controlled AC motors are provided with a range of different power inverter, variable-frequency drive or electronic commutator technologies.

The term electronic commutator is usually associated with self-commutated brushless DC motor and switched reluctance motor applications.

Major Categories

Electric motors operate on three different physical principles: magnetic, electrostatic and piezo-electric. By far the most common is magnetic.

In magnetic motors, magnetic fields are formed in both the rotor and the stator. The product between these two fields gives rise to a force, and thus a torque on the motor shaft. One, or both, of these fields must be made to change with the rotation of the motor. This is done by switching the poles on and off at the right time, or varying the strength of the pole.

The main types are DC motors and AC motors, the former increasingly being displaced by the latter.

AC electric motors are either asynchronous or synchronous.

Once started, a synchronous motor requires synchronism with the moving magnetic field's synchronous speed for all normal torque conditions.

In synchronous machines, the magnetic field must be provided by means other than induction such as from separately excited windings or permanent magnets.

A fractional horsepower (FHP) motor either has a rating below about 1 horsepower (0.746 kW), or is manufactured with a standard frame size smaller than a standard 1 HP motor. Many household and industrial motors are in the fractional horsepower class.

Major Categories by Type of Motor Commutation				
Self-Commutated			Externally Commutated	
Mechanical-Commutator Motors		Electronic-Commutator (EC) Motors	Asynchronous Machines	Synchronous Machines
AC	DC	AC	AC	
* Universal motor (AC commutator series motor or AC/DC motor) * Repulsion motor	Electrically excited DC motor: * Separately excited * Series * Shunt * Compound PM DC motor	With PM rotor: * BLDC motor With ferromagnetic rotor: * SRM	Three-phase motors: * SCIM * WRIM AC motors: * Capacitor * Resistance * Split * Shaded-pole	Three-phase motors: * WRSM * PMSM or BLAC motor - IPMSM - SPMSM * Hybrid AC motors: * Permanent-split capacitor * Hysteresis * Stepper * SyRM * SyRM-PM hybrid
Simple electronics	Rectifier, linear transistor(s) or DC chopper	More elaborate electronics	Most elaborate electronics (VFD), when provided	

Notes:

1. Rotation is independent of the frequency of the AC voltage.

2. Rotation is equal to synchronous speed (motor stator field speed).

3. In SCIM fixed-speed operation rotation is equal to synchronous speed less slip speed.

4. In non-slip energy recovery systems WRIM is usually used for motor starting but can be used to vary load speed.

5. Variable-speed operation.

6. Whereas induction and synchronous motor drives are typically with either six-step or si-

nusoidal waveform output, BLDC motor drives are usually with trapezoidal current wave-form; the behavior of both sinusoidal and trapezoidal PM machines is however identical in terms of their fundamental aspects.

7. In variable-speed operation WRIM is used in slip energy recovery and double-fed induction machine applications.

8. A cage winding is a shorted-circuited squirrel-cage rotor, a wound winding is connected externally through slip rings.

9. Mostly single-phase with some three-phase.

Abbreviations:

- BLAC - Brushless AC
- BLDC - Brushless DC
- BLDM - Brushless DC motor
- EC - Electronic commutator
- PM - Permanent magnet
- IPMSM - Interior permanent magnet synchronous motor
- PMSM - Permanent magnet synchronous motor
- SPMSM - Surface permanent magnet synchronous motor
- SCIM - Squirrel-cage induction motor
- SRM - Switched reluctance motor
- SyRM - Synchronous reluctance motor
- VFD - Variable-frequency drive
- WRIM - Wound-rotor induction motor
- WRSM - Wound-rotor synchronous motor

Self-commutated Motor

Brushed DC Motor

All self-commutated DC motors are by definition run on DC electric power. Most DC motors are small PM types. They contain a brushed internal mechanical commutation to reverse motor windings' current in synchronism with rotation.

Electrically Excited DC Motor

A commutated DC motor has a set of rotating windings wound on an armature mounted on a rotating shaft. The shaft also carries the commutator, a long-lasting rotary electrical switch that periodically reverses the flow of current in the rotor windings as the shaft rotates. Thus, every brushed

DC motor has AC flowing through its rotating windings. Current flows through one or more pairs of brushes that bear on the commutator; the brushes connect an external source of electric power to the rotating armature.

Workings of a brushed electric motor with a two-pole rotor and PM stator. ("N" and "S" designate polarities on the inside faces of the magnets; the outside faces have opposite polarities.)

The rotating armature consists of one or more coils of wire wound around a laminated, magnetically "soft" ferromagnetic core. Current from the brushes flows through the commutator and one winding of the armature, making it a temporary magnet (an electromagnet). The magnetic field produced by the armature interacts with a stationary magnetic field produced by either PMs or another winding (a field coil), as part of the motor frame. The force between the two magnetic fields tends to rotate the motor shaft. The commutator switches power to the coils as the rotor turns, keeping the magnetic poles of the rotor from ever fully aligning with the magnetic poles of the stator field, so that the rotor never stops (like a compass needle does), but rather keeps rotating as long as power is applied.

Many of the limitations of the classic commutator DC motor are due to the need for brushes to press against the commutator. This creates friction. Sparks are created by the brushes making and breaking circuits through the rotor coils as the brushes cross the insulating gaps between commutator sections. Depending on the commutator design, this may include the brushes shorting together adjacent sections – and hence coil ends – momentarily while crossing the gaps. Furthermore, the inductance of the rotor coils causes the voltage across each to rise when its circuit is opened, increasing the sparking of the brushes. This sparking limits the maximum speed of the machine, as too-rapid sparking will overheat, erode, or even melt the commutator. The current density per unit area of the brushes, in combination with their resistivity, limits the output of the motor. The making and breaking of electric contact also generates electrical noise; sparking generates RFI. Brushes eventually wear out and require replacement, and the commutator itself is subject to wear and maintenance (on larger motors) or replacement (on small motors). The commutator assembly on a large motor is a costly element, requiring precision assembly of many parts. On small motors, the commutator is usually permanently integrated into the rotor, so replacing it usually requires replacing the whole rotor.

While most commutators are cylindrical, some are flat discs consisting of several segments (typically, at least three) mounted on an insulator.

Large brushes are desired for a larger brush contact area to maximize motor output, but small brushes are desired for low mass to maximize the speed at which the motor can run without the brushes excessively bouncing and sparking. (Small brushes are also desirable for lower cost.) Stiffer brush springs can also be used to make brushes of a given mass work at a higher speed, but at the cost of greater friction losses (lower efficiency) and accelerated brush and commutator wear. Therefore, DC motor brush design entails a trade-off between output power, speed, and efficiency/wear.

DC machines are defined as follows:

- Armature circuit - A winding where the load current is carried, such that can be either stationary or rotating part of motor or generator.

- Field circuit - A set of windings that produces a magnetic field so that the electromagnetic induction can take place in electric machines.

- Commutation: A mechanical technique in which rectification can be achieved, or from which DC can be derived, in DC machines.

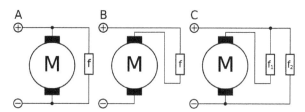

A: shunt B: series C: compound f = field coil

There are five types of brushed DC motor:-

- DC shunt-wound motor

- DC series-wound motor

- DC compound motor (two configurations):

    - Cumulative compound

    - Differentially compounded

- PM DC motor

- Separately excited

Permanent Magnet DC Motor

A PM motor does not have a field winding on the stator frame, instead relying on PMs to provide the magnetic field against which the rotor field interacts to produce torque. Compensating windings in series with the armature may be used on large motors to improve commutation under load. Because this field is fixed, it cannot be adjusted for speed control. PM fields (stators) are convenient in miniature motors to eliminate the power consumption of the field winding. Most larger DC motors are of the "dynamo" type, which have stator windings. Historically, PMs could not be made to retain high flux if they were disassembled; field windings were more practical to obtain the needed amount of flux. However, large PMs are costly, as well as dangerous and difficult to assemble; this favors wound fields for large machines.

To minimize overall weight and size, miniature PM motors may use high energy magnets made with neodymium or other strategic elements; most such are neodymium-iron-boron alloy. With their higher flux density, electric machines with high-energy PMs are at least competitive with all optimally designed singly-fed synchronous and induction electric machines. Miniature motors resemble the structure in the illustration, except that they have at least three rotor poles (to ensure starting, regardless of rotor position) and their outer housing is a steel tube that magnetically links the exteriors of the curved field magnets.

Electronic Commutator (EC) Motor

Brushless DC Motor

Some of the problems of the brushed DC motor are eliminated in the BLDC design. In this motor, the mechanical "rotating switch" or commutator is replaced by an external electronic switch synchronised to the rotor's position. BLDC motors are typically 85–90% efficient or more. Efficiency for a BLDC motor of up to 96.5% have been reported, whereas DC motors with brushgear are typically 75–80% efficient.

The BLDC motor's characteristic trapezoidal back-emf waveform is derived partly from the stator windings being evenly distributed, and partly from the placement of the rotor's PMs. Also known as electronically commutated DC or inside out DC motors, the stator windings of trapezoidal BLDC motors can be with single-phase, two-phase or three-phase and use Hall effect sensors mounted on their windings for rotor position sensing and low cost closed-loop control of the electronic commutator.

BLDC motors are commonly used where precise speed control is necessary, as in computer disk drives or in video cassette recorders, the spindles within CD, CD-ROM (etc.) drives, and mechanisms within office products such as fans, laser printers and photocopiers. They have several advantages over conventional motors:

- Compared to AC fans using shaded-pole motors, they are very efficient, running much cooler than the equivalent AC motors. This cool operation leads to much-improved life of the fan's bearings.

- Without a commutator to wear out, the life of a BLDC motor can be significantly longer compared to a DC motor using brushes and a commutator. Commutation also tends to cause a great deal of electrical and RF noise; without a commutator or brushes, a BLDC motor may be used in electrically sensitive devices like audio equipment or computers.

- The same Hall effect sensors that provide the commutation can also provide a convenient tachometer signal for closed-loop control (servo-controlled) applications. In fans, the tachometer signal can be used to derive a "fan OK" signal as well as provide running speed feedback.

- The motor can be easily synchronized to an internal or external clock, leading to precise speed control.

- BLDC motors have no chance of sparking, unlike brushed motors, making them better suited to environments with volatile chemicals and fuels. Also, sparking generates ozone which can accumulate in poorly ventilated buildings risking harm to occupants' health.

- BLDC motors are usually used in small equipment such as computers and are generally used in fans to get rid of unwanted heat.

- They are also acoustically very quiet motors which is an advantage if being used in equipment that is affected by vibrations.

Modern BLDC motors range in power from a fraction of a watt to many kilowatts. Larger BLDC motors up to about 100 kW rating are used in electric vehicles. They also find significant use in high-performance electric model aircraft.

Switched Reluctance Motor

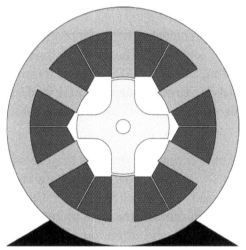

6/4 pole switched reluctance motor

The SRM has no brushes or PMs, and the rotor has no electric currents. Instead, torque comes from a slight misalignment of poles on the rotor with poles on the stator. The rotor aligns itself with the magnetic field of the stator, while the stator field windings are sequentially energized to rotate the stator field.

The magnetic flux created by the field windings follows the path of least magnetic reluctance, meaning the flux will flow through poles of the rotor that are closest to the energized poles of the stator, thereby magnetizing those poles of the rotor and creating torque. As the rotor turns, different windings will be energized, keeping the rotor turning.

SRMs are now being used in some appliances.

Universal AC-DC Motor

A commutated electrically excited series or parallel wound motor is referred to as a universal motor because it can be designed to operate on AC or DC power. A universal motor can operate well on AC because the current in both the field and the armature coils (and hence the resultant magnetic fields) will alternate (reverse polarity) in synchronism, and hence the resulting mechanical force will occur in a constant direction of rotation.

Modern low-cost universal motor, from a vacuum cleaner. Field windings are dark copper-colored, toward the back, on both sides. The rotor's laminated core is gray metallic, with dark slots for winding the coils. The commutator (partly hidden) has become dark from use; it is toward the front. The large brown molded-plastic piece in the foreground supports the brush guides and brushes (both sides), as well as the front motor bearing.

Operating at normal power line frequencies, universal motors are often found in a range less than 1000 watts. Universal motors also formed the basis of the traditional railway traction motor in electric railways. In this application, the use of AC to power a motor originally designed to run on DC would lead to efficiency losses due to eddy current heating of their magnetic components, particularly the motor field pole-pieces that, for DC, would have used solid (un-laminated) iron and they are now rarely used.

An advantage of the universal motor is that AC supplies may be used on motors which have some characteristics more common in DC motors, specifically high starting torque and very compact design if high running speeds are used. The negative aspect is the maintenance and short life problems caused by the commutator. Such motors are used in devices such as food mixers and power tools which are used only intermittently, and often have high starting-torque demands. Multiple taps on the field coil provide (imprecise) stepped speed control. Household blenders that advertise many speeds frequently combine a field coil with several taps and a diode that can be inserted in series with the motor (causing the motor to run on half-wave rectified AC). Universal motors also lend themselves to electronic speed control and, as such, are an ideal choice for devices like domestic washing machines. The motor can be used to agitate the drum (both forwards and in reverse) by switching the field winding with respect to the armature.

Whereas SCIMs cannot turn a shaft faster than allowed by the power line frequency, universal motors can run at much higher speeds. This makes them useful for appliances such as blenders, vacuum cleaners, and hair dryers where high speed and light weight are desirable. They are also commonly used in portable power tools, such as drills, sanders, circular and jig saws, where the motor's characteristics work well. Many vacuum cleaner and weed trimmer motors exceed 10,000 rpm, while many similar miniature grinders exceed 30,000 rpm.

Externally Commutated AC Machine

The design of AC induction and synchronous motors is optimized for operation on single-phase or polyphase sinusoidal or quasi-sinusoidal waveform power such as supplied for fixed-speed application from the AC power grid or for variable-speed application from VFD controllers. An AC motor has two parts: a stationary stator having coils supplied with AC to produce a rotating magnetic field, and a rotor attached to the output shaft that is given a torque by the rotating field.

Induction Motor

Large 4,500 HP AC Induction Motor.

Cage and Wound Rotor Induction Motor

An induction motor is an asynchronous AC motor where power is transferred to the rotor by electromagnetic induction, much like transformer action. An induction motor resembles a rotating transformer, because the stator (stationary part) is essentially the primary side of the transformer and the rotor (rotating part) is the secondary side. Polyphase induction motors are widely used in industry.

Induction motors may be further divided into Squirrel Cage Induction Motors and Wound Rotor Induction Motors. SCIMs have a heavy winding made up of solid bars, usually aluminum or copper, joined by rings at the ends of the rotor. When one considers only the bars and rings as a whole, they are much like an animal's rotating exercise cage, hence the name.

Currents induced into this winding provide the rotor magnetic field. The shape of the rotor bars determines the speed-torque characteristics. At low speeds, the current induced in the squirrel cage is nearly at line frequency and tends to be in the outer parts of the rotor cage. As the motor accelerates, the slip frequency becomes lower, and more current is in the interior of the winding. By shaping the bars to change the resistance of the winding portions in the interior and outer parts of the cage, effectively a variable resistance is inserted in the rotor circuit. However, the majority of such motors have uniform bars.

In a WRIM, the rotor winding is made of many turns of insulated wire and is connected to slip rings on the motor shaft. An external resistor or other control devices can be connected in the rotor circuit. Resistors allow control of the motor speed, although significant power is dissipated in the external resistance. A converter can be fed from the rotor circuit and return the slip-frequency power that would otherwise be wasted back into the power system through an inverter or separate motor-generator.

The WRIM is used primarily to start a high inertia load or a load that requires a very high starting torque across the full speed range. By correctly selecting the resistors used in the secondary resistance or slip ring starter, the motor is able to produce maximum torque at a relatively low supply current from zero speed to full speed. This type of motor also offers controllable speed.

Motor speed can be changed because the torque curve of the motor is effectively modified by the amount of resistance connected to the rotor circuit. Increasing the value of resistance will move the speed of maximum torque down. If the resistance connected to the rotor is increased beyond the point where the maximum torque occurs at zero speed, the torque will be further reduced.

When used with a load that has a torque curve that increases with speed, the motor will operate at the speed where the torque developed by the motor is equal to the load torque. Reducing the load will cause the motor to speed up, and increasing the load will cause the motor to slow down until the load and motor torque are equal. Operated in this manner, the slip losses are dissipated in the secondary resistors and can be very significant. The speed regulation and net efficiency is also very poor.

Torque Motor

A torque motor is a specialized form of electric motor which can operate indefinitely while stalled, that is, with the rotor blocked from turning, without incurring damage. In this mode of operation, the motor will apply a steady torque to the load (hence the name).

A common application of a torque motor would be the supply- and take-up reel motors in a tape drive. In this application, driven from a low voltage, the characteristics of these motors allow a relatively constant light tension to be applied to the tape whether or not the capstan is feeding tape past the tape heads. Driven from a higher voltage, (and so delivering a higher torque), the torque motors can also achieve fast-forward and rewind operation without requiring any additional mechanics such as gears or clutches. In the computer gaming world, torque motors are used in force feedback steering wheels.

Another common application is the control of the throttle of an internal combustion engine in conjunction with an electronic governor. In this usage, the motor works against a return spring to move the throttle in accordance with the output of the governor. The latter monitors engine speed by counting electrical pulses from the ignition system or from a magnetic pickup and, depending on the speed, makes small adjustments to the amount of current applied to the motor. If the engine starts to slow down relative to the desired speed, the current will be increased, the motor will develop more torque, pulling against the return spring and opening the throttle. Should the engine run too fast, the governor will reduce the current being applied to the motor, causing the return spring to pull back and close the throttle.

Synchronous Motor

A synchronous electric motor is an AC motor distinguished by a rotor spinning with coils passing magnets at the same rate as the AC and resulting magnetic field which drives it. Another way of saying this is that it has zero slip under usual operating conditions. Contrast this with an induction motor, which must slip to produce torque. One type of synchronous motor is like an induction motor except the rotor is excited by a DC field. Slip rings and brushes are used to conduct current to the rotor. The rotor poles connect to each other and move at the same speed hence the name synchronous motor. Another type, for low load torque, has flats ground onto a conventional squirrel-cage rotor to create discrete poles. Yet another, such as made by Hammond for its pre-World War II clocks, and in the older Hammond organs, has no rotor windings and discrete poles. It is not self-starting. The clock requires manual starting by a small knob on the back, while the older Hammond organs had an auxiliary starting motor connected by a spring-loaded manually operated switch.

Finally, hysteresis synchronous motors typically are (essentially) two-phase motors with a phase-shifting capacitor for one phase. They start like induction motors, but when slip rate decreases sufficiently, the rotor (a smooth cylinder) becomes temporarily magnetized. Its distributed poles make it act like a PMSM. The rotor material, like that of a common nail, will stay magnetized, but can also be demagnetized with little difficulty. Once running, the rotor poles stay in place; they do not drift.

Low-power synchronous timing motors (such as those for traditional electric clocks) may have multi-pole PM external cup rotors, and use shading coils to provide starting torque. *Telechron* clock motors have shaded poles for starting torque, and a two-spoke ring rotor that performs like a discrete two-pole rotor.

Doubly-fed Electric Machine

Doubly fed electric motors have two independent multiphase winding sets, which contribute active (i.e., working) power to the energy conversion process, with at least one of the winding sets electronically controlled for variable speed operation. Two independent multiphase winding sets (i.e., dual armature) are the maximum provided in a single package without topology duplication. Doubly-fed electric motors are machines with an effective constant torque speed range that is twice synchronous speed for a given frequency of excitation. This is twice the constant torque speed range as singly-fed electric machines, which have only one active winding set.

A doubly-fed motor allows for a smaller electronic converter but the cost of the rotor winding and slip rings may offset the saving in the power electronics components. Difficulties with controlling speed near synchronous speed limit applications.

Special Magnetic Motors

Rotary

Ironless or Coreless Rotor Motor

A miniature coreless motor

Nothing in the principle of any of the motors described above requires that the iron (steel) portions of the rotor actually rotate. If the soft magnetic material of the rotor is made in the form of a cylinder, then (except for the effect of hysteresis) torque is exerted only on the windings of the electromagnets. Taking advantage of this fact is the *coreless or ironless DC motor*, a specialized form of a PM DC motor. Optimized for rapid acceleration, these motors have a rotor that is constructed without any iron core. The rotor can take the form of a winding-filled cylinder, or a self-supporting structure comprising only the magnet wire and the bonding material. The rotor can fit inside the stator magnets; a magnetically soft stationary cylinder inside the rotor provides a return path for the stator magnetic flux. A second arrangement has the rotor winding basket surrounding the stator magnets. In that design, the rotor fits inside a magnetically soft cylinder that can serve as the housing for the motor, and likewise provides a return path for the flux.

Because the rotor is much lighter in weight (mass) than a conventional rotor formed from copper windings on steel laminations, the rotor can accelerate much more rapidly, often achieving a mechanical time constant under one ms. This is especially true if the windings use aluminum rather than the heavier copper. But because there is no metal mass in the rotor to act as a heat sink, even small coreless motors must often be cooled by forced air. Overheating might be an issue for coreless DC motor designs.

Among these types are the disc-rotor types, described in more detail in the next section.

Vibrator motors for cellular phones are sometimes tiny cylindrical PM field types, but there are also disc-shaped types which have a thin multipolar disc field magnet, and an intentionally unbalanced molded-plastic rotor structure with two bonded coreless coils. Metal brushes and a flat commutator switch power to the rotor coils.

Related limited-travel actuators have no core and a bonded coil placed between the poles of high-flux thin PMs. These are the fast head positioners for rigid-disk ("hard disk") drives. Although the contemporary design differs considerably from that of loudspeakers, it is still loosely (and incorrectly) referred to as a "voice coil" structure, because some earlier rigid-disk-drive heads moved in straight lines, and had a drive structure much like that of a loudspeaker.

Pancake or Axial Rotor Motor

A rather unusual motor design, the printed armature or pancake motor has the windings shaped as a disc running between arrays of high-flux magnets. The magnets are arranged in a circle facing the rotor with space in between to form an axial air gap. This design is commonly known as the pancake motor because of its extremely flat profile, although the technology has had many brand names since its inception, such as ServoDisc.

The printed armature (originally formed on a printed circuit board) in a printed armature motor is made from punched copper sheets that are laminated together using advanced composites to form a thin rigid disc. The printed armature has a unique construction in the brushed motor world in that it does not have a separate ring commutator. The brushes run directly on the armature surface making the whole design very compact.

An alternative manufacturing method is to use wound copper wire laid flat with a central conventional commutator, in a flower and petal shape. The windings are typically stabilized by being impregnated with electrical epoxy potting systems. These are filled epoxies that have moderate mixed viscosity and a long gel time. They are highlighted by low shrinkage and low exotherm, and are typically UL 1446 recognized as a potting compound insulated with 180 °C, Class H rating.

The unique advantage of ironless DC motors is that there is no cogging (torque variations caused by changing attraction between the iron and the magnets). Parasitic eddy currents cannot form in the rotor as it is totally ironless, although iron rotors are laminated. This can greatly improve efficiency, but variable-speed controllers must use a higher switching rate (>40 kHz) or DC because of the decreased electromagnetic induction.

These motors were originally invented to drive the capstan(s) of magnetic tape drives in the burgeoning computer industry, where minimal time to reach operating speed and minimal stopping distance were critical. Pancake motors are still widely used in high-performance servo-controlled systems, robotic systems, industrial automation and medical devices. Due to the variety of constructions now available, the technology is used in applications from high temperature military to low cost pump and basic servos.

Servo Motor

A servomotor is a motor, very often sold as a complete module, which is used within a position-control or speed-control feedback control system mainly control valves, such as motor-operated control valves. Servomotors are used in applications such as machine tools, pen plotters, and other process systems. Motors intended for use in a servomechanism must have well-documented characteristics for speed, torque, and power. The speed vs. torque curve is quite important and is high ratio for a servo motor. Dynamic response characteristics such as winding inductance and rotor inertia are also important; these factors limit the overall performance of the servomechanism loop. Large, powerful, but slow-responding servo loops may use conventional AC or DC motors and drive systems with position or speed feedback on the motor. As dynamic response requirements increase, more specialized motor designs such as coreless motors are used. AC motors' superior power density and acceleration characteristics compared to that of DC motors tends to favor PM synchronous, BLDC, induction, and SRM drive applications.

A servo system differs from some stepper motor applications in that the position feedback is continuous while the motor is running; a stepper system relies on the motor not to "miss steps" for short term accuracy, although a stepper system may include a "home" switch or other element to provide long-term stability of control. For instance, when a typical dot matrix computer printer starts up, its controller makes the print head stepper motor drive to its left-hand limit, where a position sensor defines home position and stops stepping. As long as power is on, a bidirectional counter in the printer's microprocessor keeps track of print-head position.

Stepper Motor

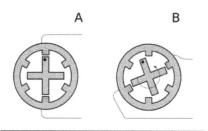

A stepper motor with a soft iron rotor, with active windings shown. In 'A' the active windings tend to hold the rotor in position. In 'B' a different set of windings are carrying a current, which generates torque and rotation.

Stepper motors are a type of motor frequently used when precise rotations are required. In a stepper motor an internal rotor containing PMs or a magnetically soft rotor with salient poles is controlled by a set of external magnets that are switched electronically. A stepper motor may also be thought of as a cross between a DC electric motor and a rotary solenoid. As each coil is energized in turn, the rotor aligns itself with the magnetic field produced by the energized field winding. Unlike a synchronous motor, in its application, the stepper motor may not rotate continuously; instead, it "steps"—starts and then quickly stops again—from one position to the next as field windings are energized and de-energized in sequence. Depending on the sequence, the rotor may turn forwards or backwards, and it may change direction, stop, speed up or slow down arbitrarily at any time.

Simple stepper motor drivers entirely energize or entirely de-energize the field windings, leading the rotor to "cog" to a limited number of positions; more sophisticated drivers can proportionally control the power to the field windings, allowing the rotors to position between the cog points and thereby rotate extremely smoothly. This mode of operation is often called microstepping. Computer controlled stepper motors are one of the most versatile forms of positioning systems, particularly when part of a digital servo-controlled system.

Stepper motors can be rotated to a specific angle in discrete steps with ease, and hence stepper motors are used for read/write head positioning in computer floppy diskette drives. They were used for the same purpose in pre-gigabyte era computer disk drives, where the precision and speed they offered was adequate for the correct positioning of the read/write head of a hard disk drive. As drive density increased, the precision and speed limitations of stepper motors made them obsolete for hard drives—the precision limitation made them unusable, and the speed limitation made them uncompetitive—thus newer hard disk drives use voice coil-based head actuator systems. (The term "voice coil" in this connection is historic; it refers to the structure in a typical (cone type) loudspeaker. This structure was used for a while to position the heads. Modern drives have a pivoted coil mount; the coil swings back and forth, something like a blade of a rotating fan. Nevertheless, like a voice coil, modern actuator coil conductors (the magnet wire) move perpendicular to the magnetic lines of force.)

Stepper motors were and still are often used in computer printers, optical scanners, and digital photocopiers to move the optical scanning element, the print head carriage (of dot matrix and inkjet printers), and the platen or feed rollers. Likewise, many computer plotters (which since the early 1990s have been replaced with large-format inkjet and laser printers) used rotary stepper motors for pen and platen movement; the typical alternatives here were either linear stepper motors or servomotors with closed-loop analog control systems.

So-called quartz analog wristwatches contain the smallest commonplace stepping motors; they have one coil, draw very little power, and have a PM rotor. The same kind of motor drives battery-powered quartz clocks. Some of these watches, such as chronographs, contain more than one stepping motor.

Closely related in design to three-phase AC synchronous motors, stepper motors and SRMs are classified as variable reluctance motor type. Stepper motors were and still are often used in computer printers, optical scanners, and computer numerical control (CNC) machines such as routers, plasma cutters and CNC lathes.

Linear Motor

A linear motor is essentially any electric motor that has been "unrolled" so that, instead of producing a torque (rotation), it produces a straight-line force along its length.

Linear motors are most commonly induction motors or stepper motors. Linear motors are commonly found in many roller-coasters where the rapid motion of the motorless railcar is controlled by the rail. They are also used in maglev trains, where the train "flies" over the ground. On a smaller scale, the 1978 era HP 7225A pen plotter used two linear stepper motors to move the pen along the X and Y axes.

Comparison by Major Categories

Comparison of motor types				
Type	Advantages	Disadvantages	Typical application	Typical drive, output
Self-commutated motors				
Brushed DC	Simple speed control Low initial cost	Maintenance (brushes) Medium lifespan Costly commutator and brushes	Steel mills Paper making machines Treadmill exercisers Automotive accessories	Rectifier, linear transistor(s) or DC chopper controller.
Brushless DC motor (BLDC) or (BLDM)	Long lifespan Low maintenance High efficiency	Higher initial cost Requires EC controller with closed-loop control	Rigid ("hard") disk drives CD/DVD players Electric vehicles RC Vehicles UAVs	Synchronous; single-phase or three-phase with PM rotor and trapezoidal stator winding; VFD typically VS PWM inverter type.
Switched reluctance motor (SRM)	Long lifespan Low maintenance High efficiency No permanent magnets Low cost Simple construction	Mechanical resonance possible High iron losses Not possible: * Open or vector control * Parallel operation Requires EC controller	Appliances Electric Vehicles Textile mills Aircraft applications	PWM and various other drive types, which tend to be used in very specialized / OEM applications.
Universal motor	High starting torque, compact, high speed.	Maintenance (brushes) Shorter lifespan Usually acoustically noisy Only small ratings are economical	Handheld power tools, blenders, vacuum cleaners, insulation blowers	Variable single phase AC, half-wave or full-wave phase-angle control with triac(s); closed-loop control optional.

AC asynchronous motors				
AC polyphase squirrel-cage or wound-rotor induction motor (SCIM) or (WRIM)	Self-starting Low cost Robust Reliable Ratings to 1+ MW Standardized types.	High starting current Lower efficiency due to need for magnetization.	Fixed-speed, traditionally, SCIM the world's workhorse especially in low performance applications of all types Variable-speed, traditionally, low-performance variable-torque pumps, fans, blowers and compressors. Variable-speed, increasingly, other high-performance constant-torque and constant-power or dynamic loads.	Fixed-speed, low performance applications of all types. Variable-speed, traditionally, WRIM drives or fixed-speed V/Hz-controlled VSDs. Variable-speed, increasingly, vector-controlled VSDs displacing DC, WRIM and single-phase AC induction motor drives.
AC SCIM split-phase capacitor-start	High power high starting torque	Speed slightly below synchronous Starting switch or relay required	Appliances Stationary Power Tools	Fixed or variable single-phase AC, variable speed being derived, typically, by full-wave phase-angle control with triac(s); closed-loop control optional.
AC SCIM split-phase capacitor-run	Moderate power High starting torque No starting switch Comparatively long life	Speed slightly below synchronous Slightly more costly	Industrial blowers Industrial machinery	
AC SCIM split-phase, auxiliary start winding	Moderate power Low starting torque	Speed slightly below synchronous Starting switch or relay required	Appliances Stationary power tools	
AC induction shaded-pole motor	Low cost Long life	Speed slightly below synchronous Low starting torque Small ratings low efficiency	Fans, appliances, record players	
AC synchronous motors				
Wound-rotor synchronous motor (WRSM)	Synchronous speed Inherently more efficient induction motor, low power factor	More costly	Industrial motors	Fixed or variable speed, three-phase; VFD typically six-step CS load-commutated inverter type or VS PWM inverter type.
Hysteresis motor	Accurate speed control Low noise No vibration High starting torque	Very low efficiency	Clocks, timers, sound producing or recording equipment, hard drive, capstan drive	Single-phase AC, two-phase capacitor-start, capacitor run motor

Synchronous reluctance motor (SyRM)	Equivalent to SCIM except more robust, more efficient, runs cooler, smaller footprint Competes with PM synchronous motor without demagnetization issues	Requires a controller Not widely available High cost	Appliances Electric vehicles Textile mills Aircraft applications	VFD can be standard DTC type or VS inverter PWM type.
Speciality motors				
Pancake or axial rotor motors	Compact design Simple speed control	Medium cost Medium lifespan	Office Equip Fans/Pumps, fast industrial and military servos	Drives can typically be brushed or brushless DC type.
Stepper motor	Precision positioning High holding torque	Some can be costly Require a controller	Positioning in printers and floppy disc drives; industrial machine tools	Not a VFD. Stepper position is determined by pulse counting.

Electromagnetism

Force and Torque

The fundamental purpose of the vast majority of the world's electric motors is to electromagnetically induce relative movement in an air gap between a stator and rotor to produce useful torque or linear force.

According to Lorentz force law the force of a winding conductor can be given simply by:

$$\mathbf{F} = I\ell \times \mathbf{B}$$

or more generally, to handle conductors with any geometry:

$$\mathbf{F} = \mathbf{J} \times \mathbf{B}$$

The most general approaches to calculating the forces in motors use tensors.

Power

Where rpm is shaft speed and T is torque, a motor's mechanical power output P_{em} is given by,

in British units with T expressed in foot-pounds,

$$P_{em} = \frac{rpm \times T}{5252} \text{ (horsepower), and,}$$

in SI units with shaft angular speed expressed in radians per second, and T expressed in newton-meters,

$$P_{em} = angularspeed \times T \text{ (watts).}$$

For a linear motor, with force F expressed in newtons and velocity v expressed in meters per second,

$$P_{em} = F \times v \text{ (watts)}.$$

In an asynchronous or induction motor, the relationship between motor speed and air gap power is, neglecting skin effect, given by the following:

$$P_{airgap} = \frac{R_r}{s} * I_r^2, \quad \text{where}$$

R_r - rotor resistance

I_r^2 - square of current induced in the rotor

s - motor slip; ie, difference between synchronous speed and slip speed, which provides the relative movement needed for current induction in the rotor.

Back Emf

Since the armature windings of a direct-current or universal motor are moving through a magnetic field, they have a voltage induced in them. This voltage tends to oppose the motor supply voltage and so is called "back electromotive force (emf)". The voltage is proportional to the running speed of the motor. The back emf of the motor, plus the voltage drop across the winding internal resistance and brushes, must equal the voltage at the brushes. This provides the fundamental mechanism of speed regulation in a DC motor. If the mechanical load increases, the motor slows down; a lower back emf results, and more current is drawn from the supply. This increased current provides the additional torque to balance the new load.

In AC machines, it is sometimes useful to consider a back emf source within the machine; this is of particular concern for close speed regulation of induction motors on VFDs, for example.

Losses

Motor losses are mainly due to resistive losses in windings, core losses and mechanical losses in bearings, and aerodynamic losses, particularly where cooling fans are present, also occur.

Losses also occur in commutation, mechanical commutators spark, and electronic commutators and also dissipate heat.

Efficiency

To calculate a motor's efficiency, the mechanical output power is divided by the electrical input power:

$$\eta = \frac{P_m}{P_e},$$

where η is energy conversion efficiency, P_e is electrical input power, and P_m is mechanical output power:

$$P_e = IV$$

$$P_m = T\omega$$

where V is input voltage, I is input current, T is output torque, and ω is output angular velocity. It is possible to derive analytically the point of maximum efficiency. It is typically at less than 1/2 the stall torque.

Various regulatory authorities in many countries have introduced and implemented legislation to encourage the manufacture and use of higher-efficiency electric motors.

Goodness Factor

Professor Eric Laithwaite proposed a metric to determine the 'goodness' of an electric motor:

$$G = \frac{\omega}{resistance \times reluctance} = \frac{\omega \mu \sigma A_m A_e}{l_m l_e}$$

Where:

- G is the goodness factor (factors above 1 are likely to be efficient)

- A_m, A_e are the cross sectional areas of the magnetic and electric circuit

- l_m, l_e are the lengths of the magnetic and electric circuits

- μ is the permeability of the core

- ω is the angular frequency the motor is driven at

From this, he showed that the most efficient motors are likely to have relatively large magnetic poles. However, the equation only directly relates to non PM motors.

Performance Parameters

Torque Capability of Motor Types

All the electromagnetic motors, and that includes the types mentioned here derive the torque from the vector product of the interacting fields. For calculating the torque it is necessary to know the fields in the air gap . Once these have been established by mathematical analysis using FEA or other tools the torque may be calculated as the integral of all the vectors of force multiplied by the radius of each vector. The current flowing in the winding is producing the fields and for a motor using a magnetic material the field is not linearly proportional to the current. This makes the calculation difficult but a computer can do the many calculations needed.

Once this is done a figure relating the current to the torque can be used as a useful parameter for motor selection. The maximum torque for a motor will depend on the maximum current although this will usually be .only usable until thermal considerations take precedence.

When optimally designed within a given core saturation constraint and for a given active current (i.e., torque current), voltage, pole-pair number, excitation frequency (i.e., synchronous speed), and air-gap flux density, all categories of electric motors or generators will exhibit virtually the same maximum continuous shaft torque (i.e., operating torque) within a given air-gap area with winding slots and back-iron depth, which determines the physical size of electromagnetic core. Some applications require bursts of torque beyond the maximum operating torque, such as short bursts of torque to accelerate an electric vehicle from standstill. Always limited by magnetic core saturation or safe operating temperature rise and voltage, the capacity for torque bursts beyond the maximum operating torque differs significantly between categories of electric motors or generators.

Capacity for bursts of torque should not be confused with field weakening capability. Field weakening allows an electric machine to operate beyond the designed frequency of excitation. Field weakening is done when the maximum speed cannot be reached by increasing the applied voltage. This applies to only motors with current controlled fields and therefore cannot be achieved with PM motors.

Electric machines without a transformer circuit topology, such as that of WRSMs or PMSMs, cannot realize bursts of torque higher than the maximum designed torque without saturating the magnetic core and rendering any increase in current as useless. Furthermore, the PM assembly of PMSMs can be irreparably damaged, if bursts of torque exceeding the maximum operating torque rating are attempted.

Electric machines with a transformer circuit topology, such as induction machines, induction doubly-fed electric machines, and induction or synchronous wound-rotor doubly-fed (WRDF) machines, exhibit very high bursts of torque because the emf-induced active current on either side of the transformer oppose each other and thus contribute nothing to the transformer coupled magnetic core flux density, which would otherwise lead to core saturation.

Electric machines that rely on induction or asynchronous principles short-circuit one port of the transformer circuit and as a result, the reactive impedance of the transformer circuit becomes dominant as slip increases, which limits the magnitude of active (i.e., real) current. Still, bursts of torque that are two to three times higher than the maximum design torque are realizable.

The brushless wound-rotor synchronous doubly-fed (BWRSDF) machine is the only electric machine with a truly dual ported transformer circuit topology (i.e., both ports independently excited with no short-circuited port). The dual ported transformer circuit topology is known to be unstable and requires a multiphase slip-ring-brush assembly to propagate limited power to the rotor winding set. If a precision means were available to instantaneously control torque angle and slip for synchronous operation during motoring or generating while simultaneously providing brushless power to the rotor winding set, the active current of the BWRSDF machine would be independent of the reactive impedance of the transformer circuit and bursts of torque significantly higher than the maximum operating torque and far beyond the practical capability of any other type of electric machine would be realizable. Torque bursts greater than eight times operating torque have been calculated.

Continuous Torque Density

The continuous torque density of conventional electric machines is determined by the size of the air-gap area and the back-iron depth, which are determined by the power rating of the armature winding set, the speed of the machine, and the achievable air-gap flux density before core saturation. Despite the high coercivity of neodymium or samarium-cobalt PMs, continuous torque density is virtually the same amongst electric machines with optimally designed armature winding sets. Continuous torque density relates to method of cooling and permissible period of operation before destruction by overheating of windings or PM damage.

Continuous Power Density

The continuous power density is determined by the product of the continuous torque density and the constant torque speed range of the electric machine.

Standards

The following are major design, manufacturing, and testing standards covering electric motors:

- American Petroleum Institute: API 541 Form-Wound Squirrel Cage Induction Motors - 375 kW (500 Horsepower) and Larger

- American Petroleum Institute: API 546 Brushless Synchronous Machines - 500 kVA and Larger

- American Petroleum Institute: API 547 General-purpose Form-Wound Squirrel Cage Induction Motors - 250 Hp and Larger

- Institute of Electrical and Electronics Engineers: IEEE Std 112 Standard Test Procedure for Polyphase Induction Motors and Generators

- Institute of Electrical and Electronics Engineers: IEEE Std 115 Guide for Test Procedures for Synchronous Machines

- Institute of Electrical and Electronics Engineers: IEEE Std 841 Standard for Petroleum and Chemical Industry - Premium Efficiency Severe Duty Totally Enclosed Fan-Cooled (TEFC) Squirrel Cage Induction Motors - Up to and Including 370 kW (500 Hp)

- International Electrotechnical Commission: IEC 60034 Rotating Electrical Machines

- International Electrotechnical Commission: IEC 60072 Dimensions and output series for rotating electrical machines

- National Electrical Manufacturers Association: MG-1 Motors and Generators

- Underwriters Laboratories: UL 1004 - Standard for Electric Motors

Non-magnetic Motors

An electrostatic motor is based on the attraction and repulsion of electric charge. Usually, electrostatic motors are the dual of conventional coil-based motors. They typically require a high-voltage power supply, although very small motors employ lower voltages. Conventional electric motors instead employ magnetic attraction and repulsion, and require high current at low voltages. In the 1750s, the first electrostatic motors were developed by Benjamin Franklin and Andrew Gordon. Today the electrostatic motor finds frequent use in micro-electro-mechanical systems (MEMS) where their drive voltages are below 100 volts, and where moving, charged plates are far easier to fabricate than coils and iron cores. Also, the molecular machinery which runs living cells is often based on linear and rotary electrostatic motors.

A piezoelectric motor or piezo motor is a type of electric motor based upon the change in shape of a piezoelectric material when an electric field is applied. Piezoelectric motors make use of the converse piezoelectric effect whereby the material produces acoustic or ultrasonic vibrations in order to produce a linear or rotary motion. In one mechanism, the elongation in a single plane is used to make a series stretches and position holds, similar to the way a caterpillar moves.

An electrically powered spacecraft propulsion system uses electric motor technology to propel spacecraft in outer space, most systems being based on electrically powering propellant to high speed, with some systems being based on electrodynamic tethers principles of propulsion to the magnetosphere.

Power Electronics

An HVDC thyristor valve tower 16.8 m tall in a hall at Baltic Cable AB in Sweden

A battery charger is an example of a piece of power electronics

Power electronics is the application of solid-state electronics to the control and conversion of electric power. It also refers to a subject of research in electronic and electrical engineering which deals with the design, control, computation and integration of nonlinear, time-varying energy-processing electronic systems with fast dynamics.

A PCs power supply is an example of a piece of power electronics, whether inside or outside of the cabinet

The first high power electronic devices were mercury-arc valves. In modern systems the conversion is performed with semiconductor switching devices such as diodes, thyristors and transistors, pioneered by R. D. Middlebrook and others beginning in the 1950s. In contrast to electronic systems concerned with transmission and processing of signals and data, in power electronics substantial amounts of electrical energy are processed. An AC/DC converter (rectifier) is the most typical power electronics device found in many consumer electronic devices, e.g. television sets, personal computers, battery chargers, etc. The power range is typically from tens of watts to several hundred watts. In industry a common application is the variable speed drive (VSD) that is used to control an induction motor. The power range of VSDs start from a few hundred watts and end at tens of megawatts.

The power conversion systems can be classified according to the type of the input and output power

- AC to DC (rectifier)
- DC to AC (inverter)
- DC to DC (DC-to-DC converter)
- AC to AC (AC-to-AC converter)

History

Power electronics started with the development of the mercury arc rectifier. Invented by Peter Cooper Hewitt in 1902, it was used to convert alternating current (AC) into direct current (DC). From the 1920s on, research continued on applying thyratrons and grid-controlled mercury arc valves to power transmission. Uno Lamm developed a mercury valve with grading electrodes making them suitable for high voltage direct current power transmission. In 1933 selenium rectifiers were invented.

In 1947 the bipolar point-contact transistor was invented by Walter H. Brattain and John Bardeen under the direction of William Shockley at Bell Labs. In 1948 Shockley's invention of the bipolar junction transistor (BJT) improved the stability and performance of transistors, and reduced costs. By the 1950s, higher power semiconductor diodes became available and started replacing vacuum tubes. In 1956 the silicon controlled rectifier (SCR) was introduced by General Electric, greatly increasing the range of power electronics applications.

By the 1960s the improved switching speed of bipolar junction transistors had allowed for high frequency DC/DC converters. In 1976 power MOSFETs became commercially available. In 1982 the Insulated Gate Bipolar Transistor (IGBT) was introduced.

Devices

The capabilities and economy of power electronics system are determined by the active devices that are available. Their characteristics and limitations are a key element in the design of power electronics systems. Formerly, the mercury arc valve, the high-vacuum and gas-filled diode thermionic rectifiers, and triggered devices such as the thyratron and ignitron were widely used in power electronics. As the ratings of solid-state devices improved in both voltage and current-handling capacity, vacuum devices have been nearly entirely replaced by solid-state devices.

Power electronic devices may be used as switches, or as amplifiers. An ideal switch is either open or closed and so dissipates no power; it withstands an applied voltage and passes no current, or passes any amount of current with no voltage drop. Semiconductor devices used as switches can approximate this ideal property and so most power electronic applications rely on switching devices on and off, which makes systems very efficient as very little power is wasted in the switch. By contrast, in the case of the amplifier, the current through the device varies continuously according to a controlled input. The voltage and current at the device terminals follow a load line, and the power dissipation inside the device is large compared with the power delivered to the load.

Several attributes dictate how devices are used. Devices such as diodes conduct when a forward voltage is applied and have no external control of the start of conduction. Power devices such as silicon controlled rectifiers and thyristors (as well as the mercury valve and thyratron) allow control of the start of conduction, but rely on periodic reversal of current flow to turn them off. Devices such as gate turn-off thyristors, BJT and MOSFET transistors provide full switching control and can be turned on or off without regard to the current flow through them. Transistor devices also allow proportional amplification, but this is rarely used for systems rated more than a few hundred watts. The control input characteristics of a device also greatly affect design; sometimes the control input is at a very high voltage with respect to ground and must be driven by an isolated source.

As efficiency is at a premium in a power electronic converter, the losses that a power electronic device generates should be as low as possible.

Devices vary in switching speed. Some diodes and thyristors are suited for relatively slow speed and are useful for power frequency switching and control; certain thyristors are useful at a few kilohertz. Devices such as MOSFETS and BJTs can switch at tens of kilohertz up to a few megahertz in power applications, but with decreasing power levels. Vacuum tube devices dominate high power (hundreds of kilowatts) at very high frequency (hundreds or thousands of megahertz) applications. Faster switching devices minimize energy lost in the transitions from on to off and back, but may create problems with radiated electromagnetic interference. Gate drive (or equivalent) circuits must be designed to supply sufficient drive current to achieve the full switching speed possible with a device. A device without sufficient drive to switch rapidly may be destroyed by excess heating.

Practical devices have non-zero voltage drop and dissipate power when on, and take some time to pass through an active region until they reach the "on" or "off" state. These losses are a significant part of the total lost power in a converter.

Power handling and dissipation of devices is also a critical factor in design. Power electronic devices may have to dissipate tens or hundreds of watts of waste heat, even switching as efficiently as possible between conducting and non-conducting states. In the switching mode, the power controlled is much larger than the power dissipated in the switch. The forward voltage drop in the conducting state translates into heat that must be dissipated. High power semiconductors require specialized heat sinks or active cooling systems to manage their junction temperature; exotic semiconductors such as silicon carbide have an advantage over straight silicon in this respect, and germanium, once the main-stay of solid-state electronics is now little used due to its unfavorable high temperature properties.

Semiconductor devices exist with ratings up to a few kilovolts in a single device. Where very high voltage must be controlled, multiple devices must be used in series, with networks to equalize voltage across all devices. Again, switching speed is a critical factor since the slowest-switching device will have to withstand a disproportionate share of the overall voltage. Mercury valves were once available with ratings to 100 kV in a single unit, simplifying their application in HVDC systems.

The current rating of a semiconductor device is limited by the heat generated within the dies and the heat developed in the resistance of the interconnecting leads. Semiconductor devices must be designed so that current is evenly distributed within the device across its internal junctions (or channels); once a "hot spot" develops, breakdown effects can rapidly destroy the device. Certain SCRs are available with current ratings to 3000 amperes in a single unit.

Solid-state Devices

Device	Description	Ratings
Diode	Uni-polar, uncontrolled, switching device used in applications such as rectification and circuit directional current control. Reverse voltage blocking device, commonly modeled as a switch in series with a voltage source, usually 0.7 VDC. The model can be enhanced to include a junction resistance, in order to accurately predict the diode voltage drop across the diode with respect to current flow.	Up to 3000 amperes and 5000 volts in a single silicon device. High voltage requires multiple series silicon devices.
Silicon-controlled rectifier (SCR)	This semi-controlled device turns on when a gate pulse is present and the anode is positive compared to the cathode. When a gate pulse is present, the device operates like a standard diode. When the anode is negative compared to the cathode, the device turns off and blocks positive or negative voltages present. The gate voltage does not allow the device to turn off.	Up to 3000 amperes, 5000 volts in a single silicon device.
Thyristor	The thyristor is a family of three-terminal devices that include SCRs, GTOs, and MCT. For most of the devices, a gate pulse turns the device on. The device turns off when the anode voltage falls below a value (relative to the cathode) determined by the device characteristics. When off, it is considered a reverse voltage blocking device.	

Gate turn-off thyristor (GTO)	The gate turn-off thyristor, unlike an SCR, can be turned on and off with a gate pulse. One issue with the device is that turn off gate voltages are usually larger and require more current than turn on levels. This turn off voltage is a negative voltage from gate to source, usually it only needs to be present for a short time, but the magnitude s on the order of 1/3 of the anode current. A snubber circuit is required in order to provide a usable switching curve for this device. Without the snubber circuit, the GTO cannot be used for turning inductive loads off. These devices, because of developments in IGCT technology are not very popular in the power electronics realm. They are considered controlled, uni-polar and bi-polar voltage blocking.	
Triac	The triac is a device that is essentially an integrated pair of phase-controlled thyristors connected in inverse-parallel on the same chip. Like an SCR, when a voltage pulse is present on the gate terminal, the device turns on. The main difference between an SCR and a Triac is that both the positive and negative cycle can be turned on independently of each other, using a positive or negative gate pulse. Similar to an SCR, once the device is turned on, the device cannot be turned off. This device is considered bi-polar and reverse voltage blocking.	
Bipolar junction transistor (BJT)	The BJT cannot be used at high power; they are slower and have more resistive losses when compared to MOSFET type devices. To carry high current, BJTs must have relatively large base currents, thus these devices have high power losses when compared to MOSFET devices. BJTs along with MOSFETs, are also considered unipolar and do not block reverse voltage very well, unless installed in pairs with protection diodes. Generally, BJTs are not utilized in power electronics switching circuits because of the I^2R losses associated with on resistance and base current requirements. BJTs have lower current gains in high power packages, thus requiring them to be set up in Darlington configurations in order to handle the currents required by power electronic circuits. Because of these multiple transistor configurations, switching times are in the hundreds of nanoseconds to microseconds. Devices have voltage ratings which max out around 1500 V and fairly high current ratings. They can also be paralleled in order to increase power handling, but must be limited to around 5 devices for current sharing.	
Power MOSFET	The main benefit of the power MOSFET is that the base current for BJT is large compared to almost zero for MOSFET gate current. Since the MOSFET is a depletion channel device, voltage, not current, is necessary to create a conduction path from drain to source. The gate does not contribute to either drain or source current. Turn on gate current is essentially zero with the only power dissipated at the gate coming during switching. Losses in MOSFETs are largely attributed to on-resistance. The calculations show a direct correlation to drain source on-resistance and the device blocking voltage rating, BV_{dss}. Switching times range from tens of nanoseconds to a few hundred microseconds, depending on the device. MOSFET drain source resistances increase as more current flows through the device. As frequencies increase the losses increase as well, making BJTs more attractive. Power MOSFETs can be paralleled in order to increase switching current and therefore overall switching power. Nominal voltages for MOSFET switching devices range from a few volts to a little over 1000 V, with currents up to about 100 A or so. Newer devices may have higher operational characteristics. MOSFET devices are not bi-directional, nor are they reverse voltage blocking. ||	

Insulated-gate bipolar transistor (IGBT)	These devices have the best characteristics of MOSFETs and BJTs. Like MOSFET devices, the insulated gate bipolar transistor has a high gate impedance, thus low gate current requirements. Like BJTs, this device has low on state voltage drop, thus low power loss across the switch in operating mode. Similar to the GTO, the IGBT can be used to block both positive and negative voltages. Operating currents are fairly high, in excess of 1500 A and switching voltage up to 3000 V. The IGBT has reduced input capacitance compared to MOSFET devices which improves the Miller feedback effect during high dv/dt turn on and turn off.	
MOS-controlled thyristor (MCT)	The MOS-controlled thyristor is thyristor like and can be triggered on or off by a pulse to the MOSFET gate. Since the input is MOS technology, there is very little current flow, allowing for very low power control signals. The device is constructed with two MOSFET inputs and a pair of BJT output stages. Input MOSFETs are configured to allow turn on control during positive and negative half cycles. The output BJTs are configured to allow for bidirectional control and low voltage reverse blocking. Some benefits to the MCT are fast switching frequencies, fairly high voltage and medium current ratings (around 100 A or so).	
Integrated gate-commutated thyristor (IGCT)	Similar to a GTO, but without the high current requirements to turn on or off the load. The IGCT can be used for quick switching with little gate current. The devices high input impedance largely because of the MOSFET gate drivers. They have low resistance outputs that don't waste power and very fast transient times that rival that of BJTs. ABB Group company has published data sheets for these devices and provided descriptions of the inner workings. The device consists of a gate, with an optically isolated input, low on resistance BJT output transistors which lead to a low voltage drop and low power loss across the device at fairly high switching voltage and current levels. An example of this new device from ABB shows how this device improves on GTO technology for switching high voltage and high current in power electronics applications. According to ABB, the IGCT devices are capable of switching in excess of 5000 VAC and 5000 A at very high frequencies, something not possible to do efficiently with GTO devices.	

DC/AC Converters (Inverters)

DC to AC converters produce an AC output waveform from a DC source. Applications include adjustable speed drives (ASD), uninterruptable power supplies (UPS), active filters, Flexible AC transmission systems (FACTS), voltage compensators, and photovoltaic generators. Topologies for these converters can be separated into two distinct categories: voltage source inverters and current source inverters. Voltage source inverters (VSIs) are named so because the independently controlled output is a voltage waveform. Similarly, current source inverters (CSIs) are distinct in that the controlled AC output is a current waveform.

Being static power converters, the DC to AC power conversion is the result of power switching devices, which are commonly fully controllable semiconductor power switches. The output waveforms are therefore made up of discrete values, producing fast transitions rather than smooth ones. The ability to produce near sinusoidal waveforms around the fundamental frequency is dictated by the modulation technique controlling when, and for how long, the power valves are on and off. Common modulation techniques include the carrier-based technique, or Pulse-width modulation, space-vector technique, and the selective-harmonic technique.

Voltage source inverters have practical uses in both single-phase and three-phase applications. Single-phase VSIs utilize half-bridge and full-bridge configurations, and are widely used for power supplies, single-phase UPSs, and elaborate high-power topologies when used in multicell configurations. Three-phase VSIs are used in applications that require sinusoidal voltage waveforms, such as ASDs, UPSs, and some types of FACTS devices such as the STATCOM. They are also used in applications where arbitrary voltages are required as in the case of active filters and voltage compensators.

Current source inverters are used to produce an AC output current from a DC current supply. This type of inverter is practical for three-phase applications in which high-quality voltage waveforms are required.

A relatively new class of inverters, called multilevel inverters, has gained widespread interest. Normal operation of CSIs and VSIs can be classified as two-level inverters, due to the fact that power switches connect to either the positive or to the negative DC bus. If more than two voltage levels were available to the inverter output terminals, the AC output could better approximate a sine wave. It is for this reason that multilevel inverters, although more complex and costly, offer higher performance.

Each inverter type differs in the DC links used, and in whether or not they require freewheeling diodes. Either can be made to operate in square-wave or pulse-width modulation (PWM) mode, depending on its intended usage. Square-wave mode offers simplicity, while PWM can be implemented several different ways and produces higher quality waveforms.

Voltage Source Inverters (VSI) feed the output inverter section from an approximately constant-voltage source.

The desired quality of the current output waveform determines which modulation technique needs to be selected for a given application. The output of a VSI is composed of discrete values. In order to obtain a smooth current waveform, the loads need to be inductive at the select harmonic frequencies. Without some sort of inductive filtering between the source and load, a capacitive load will cause the load to receive a choppy current waveform, with large and frequent current spikes.

There are three main types of VSIs:

1. Single-phase half-bridge inverter

2. Single-phase full-bridge inverter

3. Three-phase voltage source inverter

Single-phase Half-bridge Inverter

The AC input for an ASD.

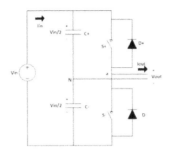

Single-Phase Half-Bridge Voltage Source Inverter

The single-phase voltage source half-bridge inverters, are meant for lower voltage applications and are commonly used in power supplies. Figure shows the circuit schematic of this inverter.

Low-order current harmonics get injected back to the source voltage by the operation of the inverter. This means that two large capacitors are needed for filtering purposes in this design. As Figure 9 illustrates, only one switch can be on at time in each leg of the inverter. If both switches in a leg were on at the same time, the DC source will be shorted out.

Inverters can use several modulation techniques to control their switching schemes. The carrier-based PWM technique compares the AC output waveform, v_c, to a carrier voltage signal, v_Δ. When v_c is greater than v_Δ, S+ is on, and when v_c is less than v_Δ, S- is on. When the AC output is at frequency fc with its amplitude at v_c, and the triangular carrier signal is at frequency f_Δ with its amplitude at v_Δ, the PWM becomes a special sinusoidal case of the carrier based PWM. This case is dubbed sinusoidal pulse-width modulation (SPWM).For this, the modulation index, or amplitude-modulation ratio, is defined as $m_a = v_c/v_\Delta$.

The normalized carrier frequency, or frequency-modulation ratio, is calculated using the equation $m_f = f_\Delta/f_c$.

If the over-modulation region, ma, exceeds one, a higher fundamental AC output voltage will be observed, but at the cost of saturation. For SPWM, the harmonics of the output waveform are at well-defined frequencies and amplitudes. This simplifies the design of the filtering components needed for the low-order current harmonic injection from the operation of the inverter. The maximum output amplitude in this mode of operation is half of the source voltage. If the maximum output amplitude, m_a, exceeds 3.24, the output waveform of the inverter becomes a square wave.

As was true for PWM, both switches in a leg for square wave modulation cannot be turned on at the same time, as this would cause a short across the voltage source. The switching scheme requires that both S+ and S- be on for a half cycle of the AC output period. The fundamental AC output amplitude is equal to $v_{o1} = v_{aN} = 2v_i/\pi$.

Its harmonics have an amplitude of $v_{oh} = v_{o1}/h$.

Therefore, the AC output voltage is not controlled by the inverter, but rather by the magnitude of the DC input voltage of the inverter.

Using selective harmonic elimination (SHE) as a modulation technique allows the switching of the inverter to selectively eliminate intrinsic harmonics. The fundamental component of the AC output voltage can also be adjusted within a desirable range. Since the AC output voltage obtained from this modulation technique has odd half and odd quarter wave symmetry, even harmonics do not exist. Any undesirable odd (N-1) intrinsic harmonics from the output waveform can be eliminated.

Single-phase Full-bridge Inverter

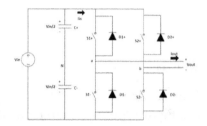

Figure: Single-Phase Voltage Source Full-Bridge Inverter

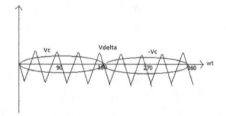

Figure: Carrier and Modulating Signals for the Bipolar Pulsewidth Modulation Technique

The full-bridge inverter is similar to the half bridge-inverter, but it has an additional leg to connect the neutral point to the load. Figure shows the circuit schematic of the single-phase voltage source full-bridge inverter.

To avoid shorting out the voltage source, S1+ and S1- cannot be on at the same time, and S2+ and S2- also cannot be on at the same time. Any modulating technique used for the full-bridge configuration should have either the top or the bottom switch of each leg on at any given time. Due to the extra leg, the maximum amplitude of the output waveform is Vi, and is twice as large as the maximum achievable output amplitude for the half-bridge configuration.

States 1 and 2 from Table 2 are used to generate the AC output voltage with bipolar SPWM. The AC output voltage can take on only two values, either Vi or −Vi. To generate these same states using a half-bridge configuration, a carrier based technique can be used. S+ being on for the half-bridge corresponds to S1+ and S2- being on for the full-bridge. Similarly, S- being on for the half-bridge corresponds to S1- and S2+ being on for the full bridge. The output voltage for this modulation technique is more or less sinusoidal, with a fundamental component that has an amplitude in the linear region of less than or equal to one $v_{o1} = v_{ab1} = v_i \cdot m_a$.

Unlike the bipolar PWM technique, the unipolar approach uses states 1, 2, 3 and 4 from Table 2 to generate its AC output voltage. Therefore, the AC output voltage can take on the values Vi, 0 or −V i. To generate these states, two sinusoidal modulating signals, Vc and −Vc, are needed, as seen in Figure.

Vc is used to generate VaN, while −Vc is used to generate VbN. The following relationship is called unipolar carrier-based SPWM $v_{o1} = 2 \cdot v_{aN1} = v_i \cdot m_a$.

The phase voltages VaN and VbN are identical, but 180 degrees out of phase with each other. The output voltage is equal to the difference of the two phase voltages, and do not contain any even harmonics. Therefore, if mf is taken, even the AC output voltage harmonics will appear at normalized odd frequencies, fh. These frequencies are centered on double the value of the normalized carrier frequency. This particular feature allows for smaller filtering components when trying to obtain a higher quality output waveform.

As was the case for the half-bridge SHE, the AC output voltage contains no even harmonics due to its odd half and odd quarter wave symmetry.

Three-phase Voltage Source Inverter

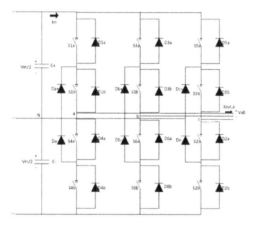

FIGURE: Three-Phase Voltage Source Inverter Circuit Schematic

FIGURE: Three-Phase Square-Wave Operation a) Switch State S1 b) Switch State S3 c) S1 Output d) S3 Output

Single-phase VSIs are used primarily for low power range applications, while three-phase VSIs cover both medium and high power range applications. Figure shows the circuit schematic for a three-phase VSI.

Switches in any of the three legs of the inverter cannot be switched off simultaneously due to this resulting in the voltages being dependent on the respective line current's polarity. States 7 and 8 produce zero AC line voltages, which result in AC line currents freewheeling through either the upper or the lower components. However, the line voltages for states 1 through 6 produce an AC line voltage consisting of the discrete values of Vi, 0 or −Vi.

For three-phase SPWM, three modulating signals that are 120 degrees out of phase with one another are used in order to produce out of phase load voltages. In order to preserve the PWM features with a single carrier signal, the normalized carrier frequency, mf, needs to be a multiple of three. This keeps the magnitude of the phase voltages identical, but out of phase with each other by 120 degrees. The maximum achievable phase voltage amplitude in the linear region, ma less than or equal to one, is $v_{phase} = v_i / 2$. The maximum achievable line voltage amplitude is $V_{ab1} = v_{ab} \cdot \sqrt{3} / 2$

The only way to control the load voltage is by changing the input DC voltage.

Current Source Inverters

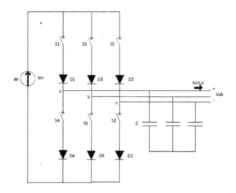

FIGURE: Three-Phase Current Source Inverter

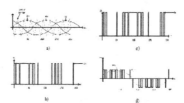

Synchronized-Pulse-Width-Modulation Waveforms for a Three-Phase Current Source Inverter a) Carrier and Modulating Signals b) S1 State c) S3 State d) Output Current

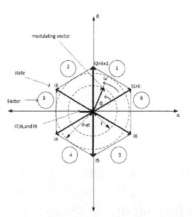

Space-Vector Representation in Current Source Inverters

Current source inverters convert DC current into an AC current waveform. In applications requiring sinusoidal AC waveforms, magnitude, frequency, and phase should all be controlled. CSIs have high changes in current overtime, so capacitors are commonly employed on the AC side, while inductors are commonly employed on the DC side. Due to the absence of freewheeling diodes, the power circuit is reduced in size and weight, and tends to be more reliable than VSIs. Although single-phase topologies are possible, three-phase CSIs are more practical.

In its most generalized form, a three-phase CSI employs the same conduction sequence as a six-pulse rectifier. At any time, only one common-cathode switch and one common-anode switch are on.

As a result, line currents take discrete values of –ii, 0 and ii. States are chosen such that a desired waveform is outputted and only valid states are used. This selection is based on modulating techniques, which include carrier-based PWM, selective harmonic elimination, and space-vector techniques.

Carrier-based techniques used for VSIs can also be implemented for CSIs, resulting in CSI line currents that behave in the same way as VSI line voltages. The digital circuit utilized for modulating signals contains a switching pulse generator, a shorting pulse generator, a shorting pulse distributor, and a switching and shorting pulse combiner. A gating signal is produced based on a carrier current and three modulating signals.

A shorting pulse is added to this signal when no top switches and no bottom switches are gated, causing the RMS currents to be equal in all legs. The same methods are utilized for each phase, however, switching variables are 120 degrees out of phase relative to one another, and the current pulses are shifted by a half-cycle with respect to output currents. If a triangular carrier is used with sinusoidal modulating signals, the CSI is said to be utilizing synchronized-pulse-width-modulation (SPWM). If full over-modulation is used in conjunction with SPWM the inverter is said to be in square-wave operation.

The second CSI modulation category, SHE is also similar to its VSI counterpart. Utilizing the gating signals developed for a VSI and a set of synchronizing sinusoidal current signals, results in symmetrically distributed shorting pulses and, therefore, symmetrical gating patterns. This allows any arbitrary number of harmonics to be eliminated. It also allows control of the fundamental line current through the proper selection of primary switching angles. Optimal switching patterns must have quarter-wave and half-wave symmetry, as well as symmetry about 30 degrees and 150 degrees. Switching patterns are never allowed between 60 degrees and 120 degrees. The current ripple can be further reduced with the use of larger output capacitors, or by increasing the number of switching pulses.

The third category, space-vector-based modulation, generates PWM load line currents that equal load line currents, on average. Valid switching states and time selections are made digitally based on space vector transformation. Modulating signals are represented as a complex vector using a transformation equation. For balanced three-phase sinusoidal signals, this vector becomes a fixed module, which rotates at a frequency, ω. These space vectors are then used to approximate the modulating signal. If the signal is between arbitrary vectors, the vectors are combined with the zero vectors I7, I8, or I9. The following equations are used to ensure that the generated currents and the current vectors are on average equivalent.

Multilevel Inverters

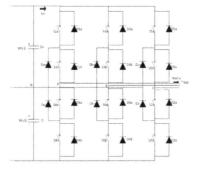

FIGURE: Three-Level Neutral-Clamped Inverter

A relatively new class called multilevel inverters has gained widespread interest. Normal operation of CSIs and VSIs can be classified as two-level inverters because the power switches connect to either the positive or the negative DC bus. If more than two voltage levels were available to the inverter output terminals, the AC output could better approximate a sine wave. For this reason multilevel inverters, although more complex and costly, offer higher performance. A three-level neutral-clamped inverter is shown in Figure.

Control methods for a three-level inverter only allow two switches of the four switches in each leg to simultaneously change conduction states. This allows smooth commutation and avoids shoot through by only selecting valid states. It may also be noted that since the DC bus voltage is shared by at least two power valves, their voltage ratings can be less than a two-level counterpart.

Carrier-based and space-vector modulation techniques are used for multilevel topologies. The methods for these techniques follow those of classic inverters, but with added complexity. Space-vector modulation offers a greater number of fixed voltage vectors to be used in approximating the modulation signal, and therefore allows more effective space vector PWM strategies to be accomplished at the cost of more elaborate algorithms. Due to added complexity and number of semiconductor devices, multilevel inverters are currently more suitable for high-power high-voltage applications. This technology reduces the harmonics hence improves overall efficiency of the scheme.

AC/AC Converters

Converting AC power to AC power allows control of the voltage, frequency, and phase of the waveform applied to a load from a supplied AC system . The two main categories that can be used to separate the types of converters are whether the frequency of the waveform is changed. AC/AC converter that don't allow the user to modify the frequencies are known as AC Voltage Controllers, or AC Regulators. AC converters that allow the user to change the frequency are simply referred to as frequency converters for AC to AC conversion. Under frequency converters there are three different types of converters that are typically used: cycloconverter, matrix converter, DC link converter (aka AC/DC/AC converter).

AC voltage controller: The purpose of an AC Voltage Controller, or AC Regulator, is to vary the RMS voltage across the load while at a constant frequency. Three control methods that are generally accepted are ON/OFF Control, Phase-Angle Control, and Pulse Width Modulation AC Chopper Control (PWM AC Chopper Control). All three of these methods can be implemented not only in single-phase circuits, but three-phase circuits as well.

ON/OFF Control: Typically used for heating loads or speed control of motors, this control method involves turning the switch on for n integral cycles and turning the switch off for m integral cycles. Because turning the switches on and off causes undesirable harmonics to be created, the switches are turned on and off during zero-voltage and zero-current conditions (zero-crossing), effectively reducing the distortion.

Phase-Angle Control: Various circuits exist to implement a phase-angle control on different waveforms, such as half-wave or full-wave voltage control. The power electronic components that are typically used are diodes, SCRs, and Triacs. With the use of these components, the user can delay the firing angle in a wave which will only cause part of the wave to be outputted.

PWM AC Chopper Control: The other two control methods often have poor harmonics, output current quality, and input power factor. In order to improve these values PWM can be used instead of the other methods. What PWM AC Chopper does is have switches that turn on and off several times within alternate half-cycles of input voltage.

Matrix converters and cycloconverters: Cycloconverters are widely used in industry for ac to ac conversion, because they are able to be used in high-power applications. They are commutated direct frequency converters that are synchronised by a supply line. The cycloconverters output voltage waveforms have complex harmonics with the higher order harmonics being filtered by the machine inductance. Causing the machine current to have fewer harmonics, while the remaining harmonics causes losses and torque pulsations. Note that in a cycloconverter, unlike other converters, there are no inductors or capacitors, i.e. no storage devices. For this reason, the instantaneous input power and the output power are equal.

Single-Phase to Single-Phase Cycloconverters: Single-Phase to Single-Phase Cycloconverters started drawing more interest recently because of the decrease in both size and price of the power electronics switches. The single-phase high frequency ac voltage can be either sinusoidal or trapezoidal. These might be zero voltage intervals for control purpose or zero voltage commutation.

Three-Phase to Single-Phase Cycloconverters: There are two kinds of three-phase to single-phase cycloconverters: 3φ to 1φ half wave cycloconverters and 3φ to 1φ bridge cycloconverters. Both positive and negative converters can generate voltage at either polarity, resulting in the positive converter only supplying positive current, and the negative converter only supplying negative current.

With recent device advances, newer forms of cycloconverters are being developed, such as matrix converters. The first change that is first noticed is that matrix converters utilize bi-directional, bipolar switches. A single phase to a single phase matrix converter consists of a matrix of 9 switches connecting the three input phases to the tree output phase. Any input phase and output phase can be connected together at any time without connecting any two switches from the same phase at the same time; otherwise this will cause a short circuit of the input phases. Matrix converters are lighter, more compact and versatile than other converter solutions. As a result, they are able to achieve higher levels of integration, higher temperature operation, broad output frequency and natural bi-directional power flow suitable to regenerate energy back to the utility.

The matrix converters are subdivided into two types: direct and indirect converters. A direct matrix converter with three-phase input and three-phase output, the switches in a matrix converter must be bi-directional, that is, they must be able to block voltages of either polarity and to conduct current in either direction. This switching strategy permits the highest possible output voltage and reduces the reactive line-side current. Therefore, the power flow through the converter is reversible. Because of its commutation problem and complex control keep it from being broadly utilized in industry.

Unlike the direct matrix converters, the indirect matrix converters has the same functionality, but uses separate input and output sections that are connected through a dc link without storage elements. The design includes a four-quadrant current source rectifier and a voltage source inverter. The input section consists of bi-directional bipolar switches. The commutation strategy can be applied by changing the switching state of the input section while the output section is in a freewheeling mode. This commutation algorithm is significantly less complexity and higher reliability as compared to a conventional direct matrix converter.

DC link converters: DC Link Converters, also referred to as AC/DC/AC converters, convert an AC input to an AC output with the use of a DC link in the middle. Meaning that the power in the converter is converted to DC from AC with the use of a rectifier, and then it is converted back to AC from DC with the use of an inverter. The end result is an output with a lower voltage and variable (higher or lower) frequency. Due to their wide area of application, the AC/DC/AC converters are the most common contemporary solution. Other advantages to AC/DC/AC converters is that they are stable in overload and no-load conditions, as well as they can be disengaged from a load without damage.

Hybrid matrix converter: Hybrid matrix converters are relatively new for AC/AC converters. These converters combine the AC/DC/AC design with the matrix converter design. Multiple types of hybrid converters have been developed in this new category, an example being a converter that uses uni-directional switches and two converter stages without the dc-link; without the capacitors or inductors needed for a dc-link, the weight and size of the converter is reduced. Two sub-categories exist from the hybrid converters, named hybrid direct matrix converter (HDMC) and hybrid indirect matrix converter (HIMC). HDMC convert the voltage and current in one stage, while the HIMC utilizes separate stages, like the AC/DC/AC converter, but without the use of an intermediate storage element.

Applications: Below is a list of common applications that each converter is used in.

- AC Voltage Controller: Lighting Control; Domestic and Industrial Heating; Speed Control of Fan,Pump or Hoist Drives, Soft Starting of Induction Motors, Static AC Switches (Temperature Control, Transformer Tap Changing, etc.)

- Cycloconverter: High-Power Low-Speed Reversible AC Motor Drives; Constant Frequency Power Supply with Variable Input Frequency; Controllable VAR Generators for Power Factor Correction; AC System Interties Linking Two Independent Power Systems.

- Matrix Converter: Currently the application of matrix converters are limited due to non-availability of bilateral monolithic switches capable of operating at high frequency, complex control law implementation, commutation and other reasons. With these developments, matrix converters could replace cycloconverters in many areas.

- DC Link: Can be used for individual or multiple load applications of machine building and construction.

Simulations of Power Electronic Systems

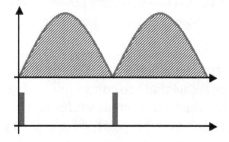

Output voltage of a full-wave rectifier with controlled thyristors

Power electronic circuits are simulated using computer simulation programs such as PLECS, PSIM and MATLAB/simulink. Circuits are simulated before they are produced to test how the circuits respond under certain conditions. Also, creating a simulation is both cheaper and faster than creating a prototype to use for testing.

Applications

Applications of power electronics range in size from a switched mode power supply in an AC adapter, battery chargers, audio amplifiers, fluorescent lamp ballasts, through variable frequency drives and DC motor drives used to operate pumps, fans, and manufacturing machinery, up to gigawatt-scale high voltage direct current power transmission systems used to interconnect electrical grids. Power electronic systems are found in virtually every electronic device. For example:

- DC/DC converters are used in most mobile devices (mobile phones, PDA etc.) to maintain the voltage at a fixed value whatever the voltage level of the battery is. These converters are also used for electronic isolation and power factor correction. A power optimizer is a type of DC/DC converter developed to maximize the energy harvest from solar photovoltaic or wind turbine systems.

- AC/DC converters (rectifiers) are used every time an electronic device is connected to the mains (computer, television etc.). These may simply change AC to DC or can also change the voltage level as part of their operation.

- AC/AC converters are used to change either the voltage level or the frequency (international power adapters, light dimmer). In power distribution networks AC/AC converters may be used to exchange power between utility frequency 50 Hz and 60 Hz power grids.

- DC/AC converters (inverters) are used primarily in UPS or renewable energy systems or emergency lighting systems. Mains power charges the DC battery. If the mains fails, an inverter produces AC electricity at mains voltage from the DC battery. Solar inverter, both smaller string and larger central inverters, as well as solar micro-inverter are used in photovoltaics as a component of a PV system.

Motor drives are found in pumps, blowers, and mill drives for textile, paper, cement and other such facilities. Drives may be used for power conversion and for motion control. For AC motors, applications include variable-frequency drives, motor soft starters and excitation systems.

In hybrid electric vehicles (HEVs), power electronics are used in two formats: series hybrid and parallel hybrid. The difference between a series hybrid and a parallel hybrid is the relationship of the electric motor to the internal combustion engine (ICE). Devices used in electric vehicles consist mostly of dc/dc converters for battery charging and dc/ac converters to power the propulsion motor. Electric trains use power electronic devices to obtain power, as well as for vector control using pulse width modulation (PWM) rectifiers. The trains obtain their power from power lines. Another new usage for power electronics is in elevator systems. These systems may use thyristors, inverters, permanent magnet motors, or various hybrid systems that incorporate PWM systems and standard motors.

Inverters

In general, inverters are utilized in applications requiring direct conversion of electrical energy from DC to AC or indirect conversion from AC to AC. DC to AC conversion is useful for many fields, including power conditioning, harmonic compensation, motor drives, and renewable energy grid-integration.

In power systems it is often desired to eliminate harmonic content found in line currents. VSIs can be used as active power filters to provide this compensation. Based on measured line currents and voltages, a control system determines reference current signals for each phase. This is fed back through an outer loop and subtracted from actual current signals to create current signals for an inner loop to the inverter. These signals then cause the inverter to generate output currents that compensate for the harmonic content. This configuration requires no real power consumption, as it is fully fed by the line; the DC link is simply a capacitor that is kept at a constant voltage by the control system. In this configuration, output currents are in phase with line voltages to produce a unity power factor. Conversely, VAR compensation is possible in a similar configuration where output currents lead line voltages to improve the overall power factor.

In facilities that require energy at all times, such as hospitals and airports, UPS systems are utilized. In a standby system, an inverter is brought online when the normally supplying grid is interrupted. Power is instantaneously drawn from onsite batteries and converted into usable AC voltage by the VSI, until grid power is restored, or until backup generators are brought online. In an online UPS system, a rectifier-DC-link-inverter is used to protect the load from transients and harmonic content. A battery in parallel with the DC-link is kept fully charged by the output in case the grid power is interrupted, while the output of the inverter is fed through a low pass filter to the load. High power quality and independence from disturbances is achieved.

Various AC motor drives have been developed for speed, torque, and position control of AC motors. These drives can be categorized as low-performance or as high-performance, based on whether they are scalar-controlled or vector-controlled, respectively. In scalar-controlled drives, fundamental stator current, or voltage frequency and amplitude, are the only controllable quantities. Therefore, these drives are employed in applications where high quality control is not required, such as fans and compressors. On the other hand, vector-controlled drives allow for instantaneous current and voltage values to be controlled continuously. This high performance is necessary for applications such as elevators and electric cars.

Inverters are also vital to many renewable energy applications. In photovoltaic purposes, the inverter, which is usually a PWM VSI, gets fed by the DC electrical energy output of a photovoltaic module or array. The inverter then converts this into an AC voltage to be interfaced with either a load or the utility grid. Inverters may also be employed in other renewable systems, such as wind turbines. In these applications, the turbine speed usually varies causing changes in voltage frequency and sometimes in the magnitude. In this case, the generated voltage can be rectified and then inverted to stabilize frequency and magnitude.

Smart Grid

A smart grid is a modernized electrical grid that uses information and communications technology to gather and act on information, such as information about the behaviors of suppliers and consumers, in an automated fashion to improve the efficiency, reliability, economics, and sustainability of the production and distribution of electricity.

Electric power generated by wind turbines and hydroelectric turbines by using induction generators can cause variances in the frequency at which power is generated. Power electronic devices are utilized in these systems to convert the generated ac voltages into high-voltage direct current (HVDC). The HVDC power can be more easily converted into three phase power that is coherent with the power associated to the existing power grid. Through these devices, the power delivered by these systems is cleaner and has a higher associated power factor. Wind power systems optimum torque is obtained either through a gearbox or direct drive technologies that can reduce the size of the power electronics device.

Electric power can be generated through photovoltaic cells by using power electronic devices. The produced power is usually then transformed by solar inverters. Inverters are divided into three different types: central, module-integrated and string. Central converters can be connected either in parallel or in series on the DC side of the system. For photovoltaic "farms", a single central converter is used for the entire system. Module-integrated converters are connected in series on either the DC or AC side. Normally several modules are used within a photovoltaic system, since the system requires these converters on both DC and AC terminals. A string converter is used in a system that utilizes photovoltaic cells that are facing different directions. It is used to convert the power generated to each string, or line, in which the photovoltaic cells are interacting.

Grid Voltage Regulation

Power electronics can be used to help utilities adapt to the rapid increase in distributed residential/commercial solar power generation. Germany and parts of Hawaii, California and New Jersey require costly studies to be conducted before approving new solar installations. Relatively small-scale ground- or pole-mounted devices create the potential for a distributed control infrastructure to monitor and manage the flow of power. Traditional electromechanical systems, such as capacitor banks or voltage regulators at substations, can take minutes to adjust voltage and can be distant from the solar installations where the problems originate. If voltage on a neighborhood circuit goes too high, it can endanger utility crews and cause damage to both utility and customer equipment. Further, a grid fault causes photovoltaic generators to shut down immediately, spiking demand for grid power. Smart grid-based regulators are more controllable than far more numerous consumer devices.

In another approach, a group of 16 western utilities called the Western Electric Industry Leaders called for mandatory use of "smart inverters". These devices convert DC to household AC and can also help with power quality. Such devices could eliminate the need for expensive utility equipment upgrades at a much lower total cost.

References

- By Jeffrey L. Rodengen. ISBN 0-945903-25-1. Published by Write Stuff Syndicate, Inc. in 1995. "The Legend of Honeywell."
- Chelikowski, J. (2004) "Introduction: Silicon in all its Forms", p. 1 in Silicon: evolution and future of a

technology. P. Siffert and E. F. Krimmel (eds.). Springer, ISBN 3-540-40546-1.

- McFarland, Grant (2006) Microprocessor design: a practical guide from design planning to manufacturing. McGraw-Hill Professional. p. 10. ISBN 0-07-145951-0.

- Heywang, W. and Zaininger, K. H. (2004) "Silicon: The Semiconductor Material", p. 36 in Silicon: evolution and future of a technology. P. Siffert and E. F. Krimmel (eds.). Springer, 2004 ISBN 3-540-40546-1.

- Price, Robert W. (2004). Roadmap to Entrepreneurial Success. AMACOM Div American Mgmt Assn. p. 42. ISBN 978-0-8144-7190-6.

- Kaplan, Daniel (2003). Hands-On Electronics. New York: Cambridge University Press. pp. 47–54, 60–61. ISBN 978-0-511-07668-8.

- Streetman, Ben (1992). Solid State Electronic Devices. Englewood Cliffs, NJ: Prentice-Hall. pp. 301–305. ISBN 0-13-822023-9.

- Horowitz, Paul; Winfield Hill (1989). The Art of Electronics (2nd ed.). Cambridge University Press. p. 115. ISBN 0-521-37095-7.

- Sedra, A.S. & Smith, K.C. (2004). Microelectronic circuits (Fifth ed.). New York: Oxford University Press. p. 397 and Figure 5.17. ISBN 0-19-514251-9.

- E. A. Marland, Early Electrical Communication, Abelard-Schuman Ltd, London 1964, no ISBN, Library of Congress 64-20875, pages 17-19;

- Ronalds, B.F. (2016). Sir Francis Ronalds: Father of the Electric Telegraph. London: Imperial College Press. ISBN 978-1-78326-917-4.

- Beauchamp, K.G. (2001). History of Telegraphy: Its Technology and Application. IET. pp. 394–395. ISBN 0-85296-792-6.

- Wilson, Arthur (1994). The Living Rock: The Story of Metals Since Earliest Times and Their Impact on Civilization. p. 203. Woodhead Publishing. ISBN 978-1-85573-301-5.

- Blundel, Stephen J. (2012). Magnetism A Very Short Introduction. Oxford University Press. p. 36. ISBN 978-0-19-960120-2.

- Garrison, Ervan G. (1998). A History of Engineering and Technology: Artful Methods (2nd ed.). CRC Press. ISBN 0-8493-9810-X. Retrieved May 7, 2009.

- Smil, Vaclav (2005). Creating the Twentieth Century:Technical Innovations of 1867–1914 and Their Lasting Impact. Oxford University Press. p. 76. ISBN 978-0-19-988341-7.

- Froehlich, Fritz E.; Kent, Allen (1 December 1998). The Froehlich/Kent Encyclopedia of Telecommunications: Volume 17 - Television Technology. CRC Press. pp. 37–. ISBN 978-0-8247-2915-8. Retrieved 10 October 2012.

- Drury, Bill (2001). Control Techniques Drives and Controls Handbook. Institution of Electrical Engineers. p. xiv. ISBN 978-0-85296-793-5.

- Langsdorf, Alexander Suss (1955). Theory of Alternating-Current Machinery (2nd ed.). Tata McGraw-Hill. p. 245. ISBN 0-07-099423-4.

- Klooster, John W. (2009). Icons of Invention: The Makers of the Modern World from Gutenberg to Gates. ABC-CLIO, LLC. p. 305. ISBN 978-0-313-34746-7. Retrieved 10 September 2012.

- Day, Lance; McNeil, Ian, eds. (1996). Biographical Dictionary of the History of Technology. London: Routledge. p. 1204. ISBN 0-203-02829-5. Retrieved 2 December 2012.

Progress of Electrical Engineering

This section focuses on the progress made by electrical engineering in the last few decades. It includes the span from the discovery of electrical current to the harnessing of electric energy. The theme of this chapter helps the readers in developing an in-depth understanding of electrical engineering.

ENIAC in Philadelphia as Glen Beck (background) and Betty Snyder (foreground) program it in BRL building 328. Photo circa. 1947 to 1955

Ancient Developments

Thales of Miletus, an ancient Greek philosopher, writing at around 600 B.C.E., described a form of static electricity, noting that rubbing fur on various substances, such as amber, would cause a particular attraction between the two. He noted that the amber buttons could attract light objects such as hair and that if they rubbed the amber for long enough they could even get a spark to jump.

At around 450 B.C.E. Democritus, a later Greek philosopher, developed an atomic theory that was remarkably similar to our modern atomic theory. His mentor, Leucippus, is credited with this same theory. The hypothesis of Leucippus and Democritus held everything to be composed of atoms. But these *atoms*, called "atomos", were indivisible, and indestructible. He presciently stated that between atoms lies empty space, and that atoms are constantly in motion. He was incorrect only in stating that atoms come in different sizes and shapes. Each object had its own shaped and sized atom.

An object found in Iraq in 1938, dated to about 250 B.C.E. and called the Baghdad Battery, resembles a galvanic cell and is believed by some to have been used for electroplating in Mesopotamia, although this has not yet been proven.

17th Century Developments

A voltaic pile, the first battery

Alessandro Volta showing the earliest pile to emperor Napoleon Bonaparte

Electricity would remain little more than an intellectual curiosity for millennia. In 1600, the English scientist, William Gilbert extended the study of Cardano on electricity and magnetism, distinguishing the lodestone effect from static electricity produced by rubbing amber. He coined the New Latin word *electricus* to refer to the property of attracting small objects after being rubbed. This association gave rise to the English words "electric" and "electricity", which made their first appearance in print in Thomas Browne's *Pseudodoxia Epidemica* of 1646.

Further work was conducted by Otto von Guericke who showed electrostatic repulsion. Robert Boyle also published work.

18th Century Developments

By 1705, Francis Hauksbee had discovered that if he placed a small amount of mercury in the glass of his modified version of Otto von Guericke's generator, evacuated the air from it to create a mild vacuum and rubbed the ball in order to build up a charge, a glow was visible if he placed his hand on the outside of the ball. This glow was bright enough to read by. It seemed to be similar to St. Elmo's Fire. This effect later became the basis of the gas-discharge lamp, which led to neon lighting and mercury vapor lamps. In 1706 he produced an 'Influence machine' to generate this effect. He was elected a Fellow of the Royal Society the same year.

Hauksbee continued to experiment with electricity, making numerous observations and developing machines to generate and demonstrate various electrical phenomena. In 1709 he published *Physico-Mechanical Experiments on Various Subjects* which summarized much of his scientific work.

Stephen Gray discovered the importance of insulators and conductors. C. F. du Fay seeing his work, developed a "two-fluid" theory of electricity.

Benjamin Franklin

In the 18th century, Benjamin Franklin conducted extensive research in electricity, selling his possessions to fund his work. In June 1752 he is reputed to have attached a metal key to the bottom of a dampened kite string and flown the kite in a storm-threatened sky. A succession of sparks jumping from the key to the back of his hand showed that lightning was indeed electrical in nature. He also explained the apparently paradoxical behavior of the Leyden jar as a device for storing large amounts of electrical charge, by coming up with the single fluid, two states theory of electricity.

In 1791, Italian Luigi Galvani published his discovery of bioelectricity, demonstrating that electricity was the medium by which nerve cells passed signals to the muscles. Alessandro Volta's battery, or voltaic pile, of 1800, made from alternating layers of zinc and copper, provided scientists with a more reliable source of electrical energy than the electrostatic machines previously used.

19th Century Developments

Electrical engineering became a profession in the late 19th century. Practitioners had created a global electric telegraph network and the first electrical engineering institutions to support the new discipline were founded in the UK and USA. Although it is impossible to precisely pinpoint a first electrical engineer, Francis Ronalds stands ahead of the field, who created the first working electric telegraph system in 1816 and documented his vision of how the world could be transformed by electricity. Over 50 years later, he joined the new Society of Telegraph Engineers (soon to be renamed the Institution of Electrical Engineers) where he was regarded by other members as the first of their cohort. The donation of his extensive electrical library was a considerable boon for the fledgling Society.

Michael Faraday portrayed by Thomas Phillips c. 1841–1842

Development of the scientific basis for electrical engineering, with the tools of modern research techniques, intensified during the 19th century. Notable developments early in this century include the work of Georg Ohm, who in 1827 quantified the relationship between the electric current and potential difference in a conductor, Michael Faraday, the discoverer of electromagnetic induction in 1831. In the 1830s, Georg Ohm also constructed an early electrostatic machine. The homopolar generator was developed first by Michael Faraday during his memorable experiments in 1831. It was the beginning of modern dynamos — that is, electrical generators which operate using a magnetic field. The invention of the industrial generator, which didn't need external magnetic power in 1866 by Werner von Siemens made a large series of other inventions in the wake possible.

In 1873 James Clerk Maxwell published a unified treatment of electricity and magnetism in A Treatise on Electricity and Magnetism which stimulated several theorists to think in terms of fields described by Maxwell's equations. In 1878, the British inventor James Wimshurst developed an apparatus that had two glass disks mounted on two shafts. It was not till 1883 that the Wimshurst machine was more fully reported to the scientific community.

Thomas Edison built the world's first large-scale electrical supply network

During the latter part of the 1800s, the study of electricity was largely considered to be a subfield of physics. It was not until the late 19th century that universities started to offer degrees in electrical engineering. In 1882, Darmstadt University of Technology founded the first chair and the first faculty of electrical engineering worldwide. In the same year, under Professor Charles Cross, the Massachusetts Institute of Technology began offering the first option of Electrical Engineering within a physics department. In 1883, Darmstadt University of Technology and Cornell University introduced the world's first courses of study in electrical engineering and in 1885 the University College London founded the first chair of electrical engineering in the United Kingdom. The University of Missouri subsequently established the first department of electrical engineering in the United States in 1886.

During this period commercial use of electricity increased dramatically. Starting in the late 1870s cities started installing large scale electric street lighting systems based on arc lamps. After the development of a practical incandescent lamp for indoor lighting, Thomas Edison switched on the world's first public electric supply utility in 1882, using what was considered a relatively safe 110 volts direct current system to supply customers. Engineering advances in the 1880s, including the invention of the transformer, led to electric utilities starting to adopting alternating current, up till then used primarily in arc lighting systems, as a distribution standard for outdoor and indoor lighting (eventually replacing direct current for such purposes). In the US there was a rivalry, primarily between a Westinghouse AC and the Edison DC system known as the "War of Currents".

George Westinghouse, American entrepreneur and engineer, financially backed the development of a practical AC power network.

"By the mid-1890s the four "Maxwell equations" were recognized as the foundation of one of the strongest and most successful theories in all of physics; they had taken their place as companions, even rivals, to Newton's laws of mechanics. The equations were by then also being put to practical use, most dramatically in the emerging new technology of radio communications, but also in the telegraph, telephone, and electric power industries." By the end of the 19th century, figures in the progress of electrical engineering were beginning to emerge.

Charles Proteus Steinmetz helped foster the development of alternating current that made possible the expansion of the electric power industry in the United States, formulating mathematical theories for engineers.

Emergence of Radio and Electronics

Charles Proteus Steinmetz circa 1915

During the development of radio, many scientists and inventors contributed to radio technology and electronics. In his classic UHF experiments of 1888, Heinrich Hertz demonstrated the existence of electromagnetic waves (radio waves) leading many inventors and scientists to try to adapt them to commercial applications, such as Guglielmo Marconi (1895) and Alexander Popov (1896).

20th Century Developments

John Fleming invented the first radio tube, the diode, in 1904.

Reginald Fessenden recognized that a continuous wave needed to be generated to make speech transmission possible, and by the end of 1906 he sent the first radio broadcast of voice. Also in 1906, Robert von Lieben and Lee De Forest independently developed the amplifier tube, called the triode. Edwin Howard Armstrong enabling technology for electronic television, in 1931.

Second World War Years

The second world war saw tremendous advances in the field of electronics; especially in radar and with the invention of the magnetron by Randall and Boot at the University of Birmingham in 1940. Radio location, radio communication and radio guidance of aircraft were all developed at this time. An early electronic computing device, Colossus was built by Tommy Flowers of the GPO to decipher the coded messages of the German Lorenz cipher machine. Also developed at this time were advanced clandestine radio transmitters and receivers for use by secret agents.

An American invention at the time was a device to scramble the telephone calls between Winston Churchill and Franklin D. Roosevelt. This was called the Green Hornet system and worked by inserting noise into the signal. The noise was then extracted at the receiving end. This system was never broken by the Germans.

A great amount of work was undertaken in the United States as part of the War Training Program in the areas of radio direction finding, pulsed linear networks, frequency modulation, vacuum tube circuits, transmission line theory and fundamentals of electromagnetic engineering. These studies were published shortly after the war in what became known as the 'Radio Communication Series' published by McGraw-Hill in 1946.

In 1941 Konrad Zuse presented the Z3, the world's first fully functional and programmable computer.

Post War Developments

Prior to the second world war the subject was commonly known as 'radio engineering' and basically was restricted to aspects of communications and radar, commercial radio and early television. At this time, study of radio engineering at universities could only be undertaken as part of a physics degree.

Later, in post war years, as consumer devices began to be developed, the field broadened to include modern TV, audio systems, Hi-Fi and latterly computers and microprocessors. In 1946 the ENIAC (Electronic Numerical Integrator and Computer) of John Presper Eckert and John Mauchly followed, beginning the computing era. The arithmetic performance of these machines allowed engineers to develop completely new technologies and achieve new objectives, including the Apollo missions and the NASA moon landing.

The invention of the transistor in 1947 by William B. Shockley, John Bardeen and Walter Brattain opened the door for more compact devices and led to the development of the integrated circuit in 1958 by Jack Kilby and independently in 1959 by Robert Noyce. In the mid to late 1950s, the term radio engineering gradually gave way to the name electronics engineering, which then became a stand alone university degree subject, usually taught alongside electrical engineering with which it had become associated due to some similarities. In 1968 Marcian Hoff invented the first microprocessor at Intel and thus ignited the development of the personal computer. The first realization of the microprocessor was the Intel 4004, a 4-bit processor developed in 1971, but only in 1973 did the Intel 8080, an 8-bit processor, make the building of the first personal computer, the Altair 8800, possible.

References

- Baigrie, Brian (2006), Electricity and Magnetism: A Historical Perspective, Greenwood Press, pp. 7–8, ISBN 0-313-33358-0

- Srodes, James (2002), Franklin: The Essential Founding Father, Regnery Publishing, pp. 92–94, ISBN 0-89526-163-4 It is uncertain if Franklin personally carried out this experiment, but it is popularly attributed to him.

- Ronalds, B.F. (2016). Sir Francis Ronalds: Father of the Electric Telegraph. London: Imperial College Press. ISBN 978-1-78326-917-4.

- Weber, Ernst; Frederik Nebeker (1994). The Evolution of Electrical Engineering: A Personal Perspective. IEEE Press. ISBN 0-7803-1066-7.

- Ronalds, B.F. (Feb 2016). "The Bicentennial of Francis Ronalds's Electric Telegraph". Physics Today. doi:10.1063/PT.3.3079.

- Ronalds, B.F. (July 2016). "Francis Ronalds (1788-1873): The First Electrical Engineer?". Proceedings of the IEEE. doi:10.1109/JPROC.2016.2571358.

- Quentin R. Skrabec, The 100 Most Significant Events in American Business: An Encyclopedia, ABC-CLIO - 2012, page 86

Permissions

All chapters in this book are published with permission under the Creative Commons Attribution Share Alike License or equivalent. Every chapter published in this book has been scrutinized by our experts. Their significance has been extensively debated. The topics covered herein carry significant information for a comprehensive understanding. They may even be implemented as practical applications or may be referred to as a beginning point for further studies.

We would like to thank the editorial team for lending their expertise to make the book truly unique. They have played a crucial role in the development of this book. Without their invaluable contributions this book wouldn't have been possible. They have made vital efforts to compile up to date information on the varied aspects of this subject to make this book a valuable addition to the collection of many professionals and students.

This book was conceptualized with the vision of imparting up-to-date and integrated information in this field. To ensure the same, a matchless editorial board was set up. Every individual on the board went through rigorous rounds of assessment to prove their worth. After which they invested a large part of their time researching and compiling the most relevant data for our readers.

The editorial board has been involved in producing this book since its inception. They have spent rigorous hours researching and exploring the diverse topics which have resulted in the successful publishing of this book. They have passed on their knowledge of decades through this book. To expedite this challenging task, the publisher supported the team at every step. A small team of assistant editors was also appointed to further simplify the editing procedure and attain best results for the readers.

Apart from the editorial board, the designing team has also invested a significant amount of their time in understanding the subject and creating the most relevant covers. They scrutinized every image to scout for the most suitable representation of the subject and create an appropriate cover for the book.

The publishing team has been an ardent support to the editorial, designing and production team. Their endless efforts to recruit the best for this project, has resulted in the accomplishment of this book. They are a veteran in the field of academics and their pool of knowledge is as vast as their experience in printing. Their expertise and guidance has proved useful at every step. Their uncompromising quality standards have made this book an exceptional effort. Their encouragement from time to time has been an inspiration for everyone.

The publisher and the editorial board hope that this book will prove to be a valuable piece of knowledge for students, practitioners and scholars across the globe.

Index

U
Utilization, 17, 20-21, 23, 165-166, 172, 175, 204

V
Voltmeter, 2, 120, 234, 257-260

W
Wired Communication, 40